SpringerBriefs in Energy

SpringerBriefs in Energy presents concise summaries of cutting-edge research and practical applications in all aspects of Energy. Featuring compact volumes of 50 to 125 pages, the series covers a range of content from professional to academic. Typical topics might include:

- A snapshot of a hot or emerging topic
- A contextual literature review
- A timely report of state-of-the art analytical techniques
- An in-depth case study
- A presentation of core concepts that students must understand in order to make independent contributions.

Briefs allow authors to present their ideas and readers to absorb them with minimal time investment.

Briefs will be published as part of Springer's eBook collection, with millions of users worldwide. In addition, Briefs will be available for individual print and electronic purchase. Briefs are characterized by fast, global electronic dissemination, standard publishing contracts, easy-to-use manuscript preparation and formatting guidelines, and expedited production schedules. We aim for publication 8–12 weeks after acceptance.

Both solicited and unsolicited manuscripts are considered for publication in this series. Briefs can also arise from the scale up of a planned chapter. Instead of simply contributing to an edited volume, the author gets an authored book with the space necessary to provide more data, fundamentals and background on the subject, methodology, future outlook, etc.

SpringerBriefs in Energy contains a distinct subseries focusing on Energy Analysis and edited by Charles Hall, State University of New York. Books for this subseries will emphasize quantitative accounting of energy use and availability, including the potential and limitations of new technologies in terms of energy returned on energy invested.

More information about this series at http://www.springer.com/series/8903

Ángel Arcos-Vargas · Laureleen Riviere

Grid Parity and Carbon Footprint

An Analysis for Residential Solar Energy in the Mediterranean Area

 Springer

Ángel Arcos-Vargas
University of Seville
Sevilla, Spain

Laureleen Riviere
University of Seville
Sevilla, Spain

ISSN 2191-5520   ISSN 2191-5539   (electronic)
SpringerBriefs in Energy
ISBN 978-3-030-06063-3   ISBN 978-3-030-06064-0   (eBook)
https://doi.org/10.1007/978-3-030-06064-0

Library of Congress Control Number: 2018964678

This Springer imprint is published by the registered company Springer Nature Switzerland AG
The registered company address is: Gewerbestrasse 11, 6330 Cham, Switzerland

Contents

Nomenclature

Capex	Capital expenditure
DHA	Time discrimination with two periods
DHS	Time discrimination with three periods
EPBC	Energy payback time
GHG	Greenhouse gases
GR	Growth rate
HP	High price scenario
INDC	Intended Nationally Determined Contribution
IRR	Internal rate of return
LCA	Life-cycle assessment
LCOE	Levelized cost of electricity
LP	Low price scenario
LR	Learning rate
NPV	Net present value
Opex	Costs of annual operations and maintenance
PR	Progress ratio
PV	Photovoltaic
REE	Red Eléctrica de España
RTE	Réseau de Transport d'Electricité
UNFCCC	United Nations Framework Convention for Climate Change
VAT	Value-added tax
WACC	Weighted average cost of capital

List of Figures

List of Tables

Abstract

In the context of global warming, big cities' atmosphere is always more contaminated and natural disasters in augmentation, solar energy, and more generally renewable energies are sources of great enthusiasm. Besides, thanks to recent improvements in technologies, the costs of photovoltaic (PV) have strongly declined in the last decades and are now accessible for particulars. The purpose of this project is then to study the economic profitability of solar energy for a residential use. A common and appropriate tool for this is the grid parity. This term, largely used in the literature, refers to the moment when producing electricity from solar modules will have the same cost than buying it from the grid.

Grid parity depends mainly on the geographic position (as solar irradiation is very different from a place to another) and on the local electricity price. Consequently, a country with expensive electricity and a high rate of irradiation is more likely to reach grid parity soon. In the present study, the geographic framework chosen is the Mediterranean area, which includes countries with similar climates but with other differences sufficiently important to obtain interesting comparative results. The PV system used in the following model is a basic one since it does not include energy storage or resale to the grid, which gives a conservative perspective to the study. Extensive financial analysis is conducted in order to determine under which conditions it is the most profitable.

The secondary objective is to evaluate the environmental impact of solar energy, mainly by carrying out carbon footprint analysis. It basically consists in comparing the emissions released by the manufacturing process of the modules to the reduction obtained thanks to its utilization. This study is first realized at an individual level and, then, is generalized at a national one in order to measure what could be the contribution of a massive investment in residential solar energy to the Paris agreement objectives.

Chapter 1
Introduction

1.1 Context and Motivation

Nowadays, the world population is growing faster than ever and the standards of living are rapidly increasing too, the combination of these two factors lead to global needs in energy higher and higher. Besides, fossil energies have never been so close to exhaustion. This brings our modern society face to a worrying problem that has to be solved as soon as possible: how will mankind fulfill its energetic needs in the next decades?

If a unique solution does not exist, for sure renewable energies have a key role to play in this challenge. This is why their market is currently going ahead so fast (the quantity of renewable energy produced within the EU-28 increased overall by 73.1% between 2004 and 2014 [1]) and many governments as well as private entities invest a lot in their development. The objective is to bring down their cost so that generating electricity from a renewable source would result in the same price than buying it from the grid. This is when grid parity will be reached and it will certainly create a new enthusiasm for these green energies.

Photovoltaic (PV) energy comes fourth, after hydropower, wind turbines and biomass, in terms of global production. Nevertheless it is the one with the highest growing rate in Europe. This is due to recent technological progress which allowed a drastic reduction of costs and a great improvement of solar modules' efficiency. As solar market knows a strong expansion, there are many things to reconsider in its organization and its structure. Especially, solar energy is becoming more affordable for particulars and governments implement all kind of financial measures to support its development. Consequently, this paper will focus on studying the feasibility and the profitability of PV systems for a domestic use.

Besides, in the common opinion, solar energy benefits from a disputed reputation. On one hand, it is considered as a green energy, totally ecological and good for the Earth, because it does not require any fossil energy and generate electricity without being supplied by any other resource. On the other hand, it is seen as a false

© The Author(s), under exclusive licence to Springer Nature Switzerland AG 2019
Á. Arcos-Vargas and L. Riviere, *Grid Parity and Carbon Footprint*,
SpringerBriefs in Energy, https://doi.org/10.1007/978-3-030-06064-0_1

green energy because of its highly contaminating components and the difficulty to recycle them at the end of their life. Moreover, people often think that solar energy is very expensive in comparison with other sources of electricity. The aim of this paper is to make the truth among these thoughts so that people wishing to invest in photovoltaic could do it being aware of all the parameters and the situation of the market.

1.2   Aims and Objectives

As previously said, the main objective of this project is to determine the financial profitability of photovoltaic systems for particulars. Thus, the study is mainly focused on whether grid parity is reached or not for solar energy. The geographic area chosen as the framework of the project is Southern Europe. Therefore, we will be able to compare the cost-effectiveness of a same PV installation under similar conditions of irradiation and under different price politics for electricity. Besides, the model selected, which will be used all along this study, is quite conservative since no energy storage nor electricity resale to the grid is considered. Solar energy is then considered here only in a context of self-consumption.

This conservative model was chosen for several reasons. First, it corresponds to the most simple to buy and install for particulars. Indeed, to stock energy, a battery and a controller have to be added to the system, and it significantly increases the costs and the complexity of the installation. Furthermore, to sell the electricity to the grid, administrative procedures to legalize the system are much more complicated. As a result, our model stands for private individuals who want to make a first and easy investment in solar energy with the aim of saving money on his electrical bills and of participating to a greener consumption. The second reason is that taking on a conservative model is like considering the worst scenario. It means that if our model is profitable, any other model would be so as well and it would demonstrate that the economic profitability of solar energy is now something established.

The second objective of the project is to deal with the ecological issue. Solar modules' fabrication requires energy and releases carbon emissions but, then, during their whole functioning life they allow to reduce carbon emissions. Therefore the aim is to compare these two amounts so as to quantify the environmental impact of solar energy.

The paper is structured in a logical way. After a literature review on grid parity which allows to understand what is the current situation of solar market and what are the scientists' forecasts about it, we will present the details of our model and all the hypothesis that are made. The model is first applied to Spain, in a section where all the methods of calculation are explained. Then, the same model is implemented in other Mediterranean countries and a comparative study of the results is displayed. An extensive financial analysis is also made for all the countries. The last chapter deals with environmental problematic, evaluing the PV system's carbon footprint and carrying out carbon balance studies. Finally, in the appendix, you will find a paper that was redacted in association with this project. This paper is focused on the environmental issue since it was sent for publication to the review *Journal of Cleaner Production* (impact factor 4.6 in 2015).

Reference

1. Eurostat Statistic Explained. http://ec.europa.eu/eurostat/statistics-explained/images/2/29/ Electricity_and_gas_prices%2C_second_half_of_year%2C_2013%E2%80%9315_%28EUR_ per_kWh%29_YB16.png

Chapter 2
Literature Review

Grid parity can be basically defined as the intersection of the price of the electricity generated by a PV system and the price of conventional electricity production (Hurtado Munoz et al. (2013) [1]). This expression was first used in a scientific publication in 2005, when an article for the magazine "Frontiers, the BP magazine of technology and innovation" related it with making solar PV competitive [1]. Since then the term has been used in almost every paper dealing with the development or the future of PV energy. This chapter will provide a literature review about the different definitions that exist, the methods of calculation and the geographic areas where it will be reached.

2.1 Different Definitions of Grid Parity

Hurtado Munoz et al. (2013) [1] aim at warning about the sometimes abusive use of the grid parity expression. Indeed, too many articles present results of grid parity estimations without specifying their methods of calculations or on which electricity prices they have based their study. In the same goal of being more precise about grid parity, Esram et al. [2] determines four different types of grid parity that are featured in Table 2.1.

The first row of Table 2.1 indicates for each kind of parity with which electricity rate will be compared the delivered cost of PV electricity. We see that for the retail parity the average retail rate is used, it means that retail parity is the parity typically employed. Two major problems can be seen with this basic definition:

- The average retail rate is not always reflective of the production costs
- As PV energy is not constant and is very often used to offset peak generation, it can be a lack of sense to compare it with the average retail price.

That is the purpose of the peak parity, which takes into account the cost of fulfilling the demand during peak hours. The price of peaking hours many times

Á. Arcos-Vargas and L. Riviere, *Grid Parity and Carbon Footprint*, SpringerBriefs in Energy, https://doi.org/10.1007/978-3-030-06064-0_2

Table 2.1 Different types of parity (own elaboration)

	Spot market parity	Peak parity	Retail parity	Cost parity
Electricity rate considered	Locational marginal price at the point of end use	Generation costs from conventional peaking devices such as diesel generators	Average retail rate	Wholesale rate
Advantages	The easiest form of grid parity to reach	Peaking costs many times almost double the average retail rate	Usual definition of grid parity	It allows solar energy to compete effectively with other resources
Where it could be used	Locations where power congestion limits flows and keeps prices high	Regions where PV energy systems are meant to offset peaking units		Regions where the difference between retail and wholesale rates is high

Source Esram et al. [2]

reaches the double of the average price. That's why peak parity is the easiest to reach, but probably as well the most interesting for the solar market.

Then there are the spot market parity and the cost parity that both take into account, though in a different way, the difference between end user electricity price and production costs. It may be interesting to do so in some cases where the grid structure is complex or not well maintained since it creates high losses or high costs of transport.

As a consequence, to analyze grid parity results, it is very important to know on which electricity rates is taken into account. We will see that in the main part of the papers that are mentioned in this literature review the average retail price is used. The electricity rate chosen is not the only parameter that enters in the calculation of grid parity which relies on a quite complex model, which is presented in the next section.

2.2 Calculation of Grid Parity

The dynamics for grid parity are based on both the decline of PV electricity generation costs and the electricity prices showing an increasing path over time (even if in many analyses the retail electricity price is considered constant over time). This situation is illustrated by Bronski et al. in their report The economics of grid defection (2014) [3]: in which it is analysed that the evolution of the leveled cost of electricity (LCOE), associated with the technological improvement and reduction of costs of PV and storage facilities, as compared to the foreseeable evolution of electricity tariffs, for each US state. For the most favourable cases (e.g. Honolulu–Hawaii) this parity is reached in 2022.

The LCOE is defined as the cost that, if assigned to every unit of energy produced by the system over the lifetime period, will equal the total lifetime cost, when discounted back to the base year (Biondi and Moretto (2015), [4]). The same authors describe it as an easy tool used to compare the unit costs of different power generation technologies, along their economic lifetime. It allows to capture capital costs, on-going system-related costs and fuel costs—along with the amount of electricity produced—and converts them into a common metric. To study the LCOE of PV energy the metric used will be €/kWh, so that it can be directly compared to the electricity retail rate.

The PV energy costs have greatly decreased over the past decades as it has been analyzed by Hurtado Munoz [1] and Breyer and Gerlach [5, 6]. This decline was possible thanks to the technological progress made in the sector. This progress is measured by the learning rate (LR) or the progress ratio (PR). That's why the dynamic grid parity model is founded on the application of learning curves to the LCOE.

The LCOE can be expressed in different ways but it always includes the following parameters:

– Progress ratio (or learning rate as PR = 1 – LR)
– Capital expenditures, that is to say the total cost of buying and installing the PV system
– Annual operations and maintenance expenditures
– Growth rate of the PV industry
– An annuity factor
– The total of energy produced

We will detail here two different ways of calculating it. The first one is described by Breyer and Gerlach (2010) [5] as follow:

$$LCOE = \frac{Capex * crf + Opex}{E_{net}} \tag{2.1}$$

Actually Eq. 2.1 is just the expression of the total annual cost of a PV system divided by the total amount of energy it produces during one year (E_{net}). The *Capex* correspond to the capital expenditures and depend on the progress ratio *(PR)* and on the cumulated output level (P_x). *Crf* is the annuity factor which aims at actualizing each year the capital expenditures, its value is function of the WACC (weighted average cost of capital) and of the annual cost of insurance (k_{ins}). Finally Opex represent the costs of annual operations and maintenance, they are expressed by a percentage of the initial Capex, and are quite low (1.5% of Capex according to [5]). Equations 2.2–2.4 detail the calculation of each term.

$$Capex = c_0 * \left(\frac{P_x}{P_0}\right)^{\frac{\log (PR)}{\log 2}} \tag{2.2}$$

$$P_x = P_0 * \prod_{t=0}^{x} (1 + GR_t) \qquad (2.3)$$

$$crf = \frac{wacc * (1 + wacc)^N}{(1 + wacc)^N - 1} + k_{ins} \qquad (2.4)$$

GR *growth rate*
P_0 *initial output level*
c_0 *cost at initial output level*

Biondi and Moretto (2015) [4] approach the problem from another angle but the final purpose is the same. They express the grid parity problem with Eq. 2.5 (this point is equivalent to the method previously presented), which is equal to find t* that fulfills Eq. 2.6 (this is where it takes a different manner of approaching the problem).

$$E_t(P_{t^*}) = E_t(LCOE_{t^*}) \qquad (2.5)$$

$$t^* = \max \left[\frac{\ln\left(\frac{LCOE_{2011}}{P_{2011}}\right)}{\alpha_p - \alpha_c} \right] \qquad (2.6)$$

where $E_t(.)$ stands for the expectation value taken at the starting point t, P_t for the electricity price at t, and $LCOE_t$ for the value of the levelised cost of electricity at t.

We have already seen a major difference between both models, the second one is considering a variable price of electricity over the lifetime of the PV system whereas the first one was considering it constant. We are not going to detail how they evaluate this inconstant variable, as this is a quite complex process and studying the variation of electricity price is not the main purpose of this paper, but we are just going to point out that α_p, which is called the drift term, is the factor that takes into account the retail price variations.

α_c is the factor that represents the prediction of LCOE's dynamics. The authors have chosen an empirical-based methodology designed to describe the law of cost reduction for a specific industry. It is based on the assumption that at each doubling of cumulated capacity the unit cost decreases by a stable percentage called learning rate (LR). Additionally the growth rate (GR) has to be introduced in the calculation, they then obtain the following formula:

$$\alpha_c = \frac{\ln(1 - LR)}{\ln 2} * GR \qquad (2.7)$$

Finally, the expected value of LCOE at time t is evaluated with the bellowed expression.

$$LCOE_t = LCOE_{t_0} * e^{\alpha_c * t} \qquad (2.8)$$

where $LCOE_{t_0}$ is the value of LCOE at the starting point of the study. As we can see in Eq. 2.6 the starting point is actually 2011.

And the initial value of LCOE is, as in the precedent model, the sum of the different costs divided by the total of energy produced. The unique difference is that they consider the insurance as a direct cost:

$$LCOE = \frac{Capex + Assurance + Opex}{E_{net}} \qquad (2.9)$$

The next point we are going to focus on is the evaluation of the progress ratio and the growth rate which are two parameters that have a great influence on the results. It seems to exist a quite large consensus on the method to calculate the progress ratio and the value obtained. For instance, the three papers [1, 4, 5] use the same log-linear model.

This means that it exists a constant learning rate for PV industry. Designing a pessimistic line on this graph conducts to a learning rate of 19.3%, whereas the optimistic line gives a learning rate of 22.8%. As a consequence, it is generally admitted that the PV industry has a learning rate of 20%.

The growth rate is more complex to determine and its value differs in a significant way in function of the publications and in function of the geographic area where the model will be applied. Breyer and Gerlach (2010) [5] aim at evaluating grid parity in countries from the entire world, and for that, they need to elect a global growth rate. In that purpose, they plotted the evolution of the global PV production since the 50s and this curve presented a high and constant growth rate of 45% between 1995 and 2010. Nevertheless, it would be too optimistic to consider such a growth rate for an entire grid parity model. Consequently, in conjunction with many scientific researchers and financial analysts [7] they finally use a 30% growth rate.

Biondi and Moretto (2015) [4] apply their model only to the Italian market, therefore they have to determine PV industry's growth rate for Italy. They consider two scenarios, an optimistic one supported by the Solar energy report 2012 from the Politecnico de Milano [8] that predict an 18% growth rate. A more conservative scenario is based on previsions from the Global Market Outlook [9] and includes a 10% growth rate. The influence of the growth rate in the model is not so important, actually the difference between the results from one scenario or another is just one year and a half: in the conservative scenario Italy is supposed to reach grid parity for residential use in October 2017, and in April 2016 with the optimistic scenario.

Generally, the European market has a growth rate a little bit lower than the world market [9]. The predictions for the global market give an annual growth rate above 19% according to the medium scenario, whereas for the European market this rate hardly reach 10%.

Table 2.2 2016–2020 predictions for annual growth in a few countries of the world according to Global market Outlook for solar power 2016–2020 (own elaboration)

	2015 total capacity (MW)	Total capacity medium scenario by 2020 (MW)	New capacity (MW)	Compound annual growth rate (%)
China	43,381	130,381	87,000	25
USA	2591	8531	59,400	27
India	5048	57,398	52,350	63
Japan	34,347	63,347	29,000	13
Brazil	69	6509	6440	149
Egypt	16	4859	4843	214
Chile	854	4509	3210	26

Source Solar power Europe [8]

However this global growth rate is an average and is not representative of the disparities that exist between one country and another. Table 2.2 features the 2016–2020 prospections for the growth rate and the new capacity installed for a few countries around the world.

This table figures out that a lot of countries have a growth rate greatly superior to the average global one. It is interesting to note that even if China doesn't have the highest annual growth rate, in terms of total or new capacity installed it has a level way superior to any other nation. That's why China is now considered as the world leader of PV energy. Countries like Brazil or Egypt have growth rates remarkably high because they are just entering the PV market. The Japanese case is also relevant, as it was historically the first country to develop solar energy and it was the leader country until 2003–2004 [10] before knowing a decline mainly due to political changes. Nowadays it still appears among the leaders of the market, but as its growth rate is quite low it will soon be overpassed by many countries.

To conclude this section about the calculation of grid parity, we are going to have a look at Fig. 2.1 that features the decrease of PV modules prices overtime.

We can see that during the first years of commercialization prices are really dropping, the cost production, expressed in dollar per Watt installed, is losing more than 4$ each year. This is representative of the huge technological progress that are made when a new technology is put on the market. Then, since 1985 and until the beginning of our decade, the decline is lighter but still quite constant. Indeed, the curve stays reasonably close to a straight line with a 0.6 slope. However, as the PV module price begins to be very low, this decline will probably not follow this way forever, it will probably reach a threshold in the next few years, and PV sector will need to focus on other aspect to keep improving its competitiveness.

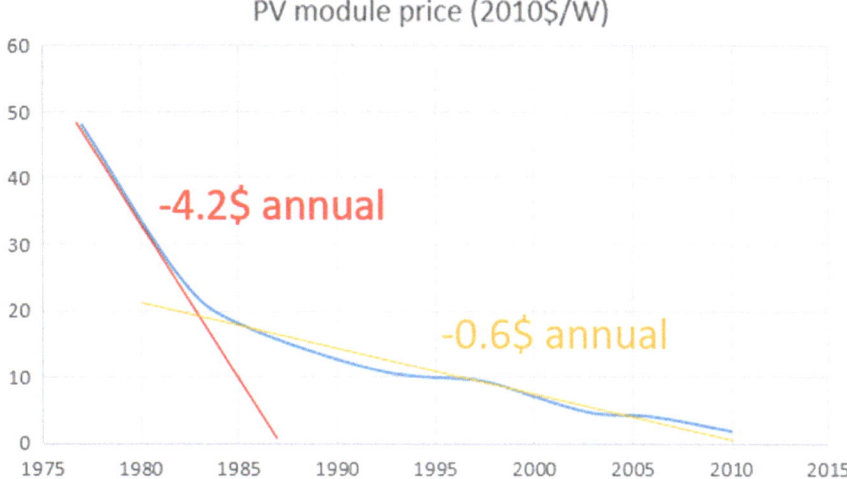

Fig. 2.1 Evolution of the PV global costs. *Source* Own elaboration, based on Breyer and Gerlach (2010) [5]

2.3   Grid Parity Achievement

We have seen that it exists many different ways of evaluating grid parity as there are different mathematical models, different kinds of parity, and the parameters' values are difficult to determine. Consequently we can't be surprised to find grid parity results that differ greatly from one paper to another. Table 2.3 synthesizes some results from 7 papers.

It highlights that Breyer and Gerlach's model [5], which was previously detailed, is very optimistic. This may come from both the relatively high growth rate they have chosen (30%) and the fact that they have considered the electricity retail price constant over the frame time of the study. According to their model, grid parity should already be there since 2010 in Spain, Portugal and Italy. And this is without taking into account the Feed-in-tariffs (Fits) that exist or existed in those countries. Another example of its optimism is that it considers for Eastern US that grid parity will be there in 2016, whereas the base case scenario of paper [3] evaluates it for the same area at 2049. However this ultimate model takes into consideration a battery system linked to the PV module. This means that for residential uses, the customers would be able to be totally self-dependent in electricity as the loaded batteries would substitute the grid when there is no solar energy available.

Moreover, an interesting point from the economics of grid defection [3] is that it presents four possible scenarios. In Table 2.3 only the two extreme scenarios were reported: the base case—which represents a conservative view on incremental progress with existing solar PV and batteries technologies as it assumes that there will not be radical improvements in technology performance or costs—and the combined improvement case—which considers both effects from accelerated

Table 2.3 Grid parity achievement for the residential sector. Compilation from seven papers

	2010	2011	2012	2013	2014	2015	2016	2017	2018	2019	2020	2021	2022	2023	2024	2025	2026	2027	2028	2029	2030
Brazil			←—— 15 ——→ 4✓																		
Germany			←4-5*→											←5**→							
Netherlands		16✓																			
Cyprus			16✓																		
Malaysia																				17✓	
Italy	4✓				3✓																
Portugal	4✓																				
Spain	4✓																				
France						4✓															
US East							4✓			10*✓											10* = 2049
US North West												4-10**✓									10* = 2037
Japan			4✓																		
Hawaï					10*✓								10✓								

> All numbers in the table correspond to the paper reference as it is used in the rest of the study.

> All the papers refer to the retail parity that will be studied hereafter, except paper 5 which considers as well a scenario of cost parity.

[5] *end user price = generation price => retail parity
**end user price = wholesale price => cost parity

[10] * base case
** combined improvement case => it is supposed that improvements will be made both in technology and management of the energy

technology improvement and demand side improvement. Consequently this scenario is based at the same time on more aggressive projections of battery prices and total installed PV costs, and on a higher flexibility to shift the load profile for the user. In Hawaï, due to its high level of solar radiation and to the elevated cost of electricity, the difference between the two scenarios is not so important (2015 for the base case scenario and 2022 for the other). But for Eastern or Western US the difference is really significant as it differs respectively of 29 and 17 years.

Bhandari and Stadler [6] have as well decided to consider different cases for their grid parity results. They distinguish between what Esram et al. [2] call cost parity and retail parity. The first one is the most difficult to reach and corresponds to the case of a customer having a PV system on his roof and aiming at selling its electrical production to the grid. Indeed, in this case, he would be paid at the wholesale rate. The second parity is the basic one where a customer uses his own electric production and hence saves the amount of money corresponding to buy this electricity to the grid. For Germany and according to this study [6], there is a 10 years gap between these both parities.

Another interesting way of featuring the grid parity is using a map like Poortmans and Sinke (2008) [11] did. LCOE depends mainly on the solar irradiation of the area and on the system's costs which can vary greatly from a continent to another due to taxes. However, these costs can certainly be considered constant all over Europe. That's why Poortmans and Sinke [11] have designed European maps where grid parity areas depend only on solar irradiation and are delimited by the purple line that is going up north over time.

According to this model, in 2015 only the South of Europe has reached grid parity, in 2020 it has already spread to the main part of Europe and in 2030 all Europe is under grid parity including the Scandinavian countries which receive very few solar irradiation.

We saw that grid parity is a concept widely used in the PV field but also that is quite complex to determine since the estimations differ a lot from a model to another. Consequently, we can wonder about the total relevance of this term. In that way, Hurtado Munoz et al. [1] lead an interesting study pointing out that "the concept of grid parity has come to be contested by a growing number of actors who criticize its relevance" and trying to "understand the role of the grid parity debate in the PV field".

Hurtado Munoz et al. [1] confirm that there is no consensus about when and where grid parity will be reached first, although it points out that a majority agree on Italy to be first. It also warns about considering grid parity as the major milestone for the PV industry, as this is not the unique problem to solve like Breyer, Gerlach and Werner [12] highlights: "Policymakers […] must recognize that dropping costs in solar technology will not automatically resolve our energy problems. If policymakers wish to help distributed solar technologies across the chasm into commercialization, political mandates to further encourage their adoption would be necessary". Moreover, despite its relatively high rate, the growth of the industry is not global and the development of the industry is limited to a certain number of countries. PV sector can't be content with this unequal growth if it wants to reach its ambitious targets.

Finally, Hurtado Munoz et al. [1] question a remarkable point: what is next to grid parity? This question raises several problems. First, the grid might not be prepared to the sudden growth that it will generate. Indeed, increased demand will result in problems with grid capacity, as the intermittent nature of distributed PV will create congestion and overloading in transmission or distribution lines. Secondly, it will cause a high demand in raw materials, which could itself engenders an increase of the prices or a lack of availability.

A major problem of PV systems for residential use is that only a low percentage of the electricity produced can be directly consumed. As a matter of fact, the daily load curve is not similar to the energy production curve because the peak demand corresponds to the early morning and to the evening when people needs light, electricity for cooking, washing, heating… while the peak production is during midday when sunlight is strongest. Nuno [13] estimates that the direct consumption is only 30% of the total, value that is coherent with the 40% evaluated by Bhandari and Stadler [6] who have centered their study in the area of Cologne (Germany).

In addition, when the production is higher than the consumption there is only two possible solutions for the non-used electricity: or supply it to the grid or store it. As we saw that the wholesale rate is superior to the retail rate, customers have interest to increase their self-consumption and to minimize the amount of electricity released to the grid. And this is where energy storage plays a key role: it permits to significantly raise the percentage of self-consumption. In a real time experiment leaded by Castillo-Cagigal et al. [14], it is demonstrated that using a 5.4 kWh battery as a storage system makes the coefficient of self-consumption go up from 32.2 to 70.5%. This is the proof that development and progress of energy storage will play a determinant role in the future of PV energy and grid parity.

References

1. Hurtado Munoz LA, Huijben JCCM, Verhees B, Verbong GPJ (2014) The power of grid parity: a discursive approach. Technol Forecast Soc Change 87:179–190
2. Esram T, Krein PT, Kuhn BT, Balog RS, Chapman PL (2008). Power electronics needs for achieving grid-parity. In: Solar energy costs, IEEE energy 2030 Atlanta, Georgia, USA, 17–18 Nov 2008
3. Bronski P, Creyts J, Guccione L, Madrazo M, Mandel J, Rader B, Seif D, Lilienthal P, Glassmire J, Abromowitz J, Crowdis M, Richardson J, Schmitt E, Tocco H (2014) The economics of grid defection
4. Biondi T, Moretto M (2015) Solar grid parity dynamics in Italy: a real option approach. Energy 80:293–302
5. Breyer C, Gerlach A (2010) Global overview on grid-parity. Prog Photovoltaics Res Appl 21:121–136
6. Bhandari R, Stadler I (2009) Grid parity analysis of solar photovoltaic systems in Germany using experience curves. Sol Energy 83:1634–1644
7. Breyer C, Gerlach A (2010) Global overview on grid-parity event dynamics. In: 25th European photovoltaic solar energy conference/WCPEC-5, Valencia, Sept 2010
8. Politecnico de milano. Solar energy report 2012. Il sistema industriale italiano nel business dell'energia solare
9. Global market Outlook for solar power 2016–2020 from Solar Power Europe
10. Lopez-Polo A, Haas R, Panzer C, Auer H (2012) Prospects for grid-parity of photovoltaics due to effective promotion schemes in major countries. Energy Economics Group, Vienna University of Technology
11. Poortmans J, Sinke W (2008) The strategic research agenda of the European PV technology platform. In: IEA: energy technology roadmap workshop, PV ERA NET 3rd joint workshop, Paris-France, 2008
12. Breyer C, Gerlach A, Werner C (2011) Grid parity: coming sooner than you think. Future Photovoltaics
13. Nuno F (2016) A regulatory frameworks for PV prosumers
14. Castillo-Cagigal M, Caamaño-Martın E, Matallanas E, Masa-Bote D, Gutierrez A, Monasterio-Huelin F, Jimenez-Leube J (2011) PV self-consumption optimization with storage and active DSM for the residential sector

Chapter 3
Model for Spain

3.1 Problem Overview

Thanks to the literature review we have seen that grid parity is a phenomenon which had been widely studied by the scientific community during the last decade. Some affirms that it is already present in a large number of countries, others are more conservative and think it will take a few years more to be reached. Grid parity depends on many factors and this is why it is quite complex to evaluate it with total accuracy. The purpose of this study is to focus on Southern Europe and, by designing a simple model, to determine which countries are already under grid parity. For those which are not, we will develop a forecast model which will give us an approximation of the time necessary for grid parity to be reached. Our model will correspond to the retail parity for a residential use. It means we will study the case of a typical household that is wishing to install solar panels in order to save money on their electrical bill. The PV system will not totally replaced the connection to the grid, it will act as an alternative to the grid during sunned hours and it will not include energy storage nor resale to the network (see Fig. 3.1).

The PV system incorporated in the model will be the same for each country. The variable parameters between one country and another are the price of electricity and the solar irradiation. In this chapter, we will first explain how we determine these two parameters. Then, the PV system will be presented in detail with all its components and their respective costs. Finally, to conclude the chapter, the resolution of the model for the Spanish case will be displayed. Results for other countries will be revealed in Chap. 4.

Á. Arcos-Vargas and L. Riviere, *Grid Parity and Carbon Footprint*, SpringerBriefs in Energy, https://doi.org/10.1007/978-3-030-06064-0_3

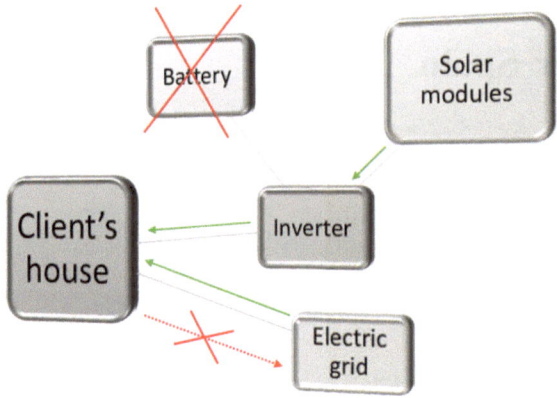

Fig. 3.1 Diagram of the whole PV system. *Source* Own elaboration

3.1.1 Electricity Prices

3.1.1.1 Electricity Retail Price Within Europe

Electricity prices will play an important role in our study. Indeed, a country with a high retail price will be more likely to reach grid parity than another where consumers already have access to cheap electricity. Figure 3.2 shows 2015 electricity prices for most European countries [1], they are retail prices, which means it corresponds to the price the domestic final consumer will pay for using 1 kWh.

The first striking point is that these prices are not homogeneous at all, the higher price is five times bigger than the cheaper one. Then, there is a logical split between Eastern Europe with lower prices and Western Europe with higher prices, which is in harmony with the difference of general cost of life in these both areas. However, even within countries with the same standard of living, electricity prices can differ

Fig. 3.2 Domestic electricity price within Europe (2015). *Source* Eurostat Statistic Explained [1] and own elaboration

greatly. For instance, in Germany or in Denmark it is almost two times higher than in France. The principal reason for this is the taxes policy of each country. Actually, in Germany and in Denmark, taxes represent respectively the third and the half of the total retail price [1], and if we consider prices without taxes they are not above the European average. The second reason that explains such differences of prices is the way electricity is produced. In France, approximately 70% (in function of the months) of the national electric production comes from nuclear plants, which allows a cheap production. On the contrary, islands like Cyprus or Malta are totally dependent on fuel and importation. Their electricity price before taxes is consequently relatively high, though it is compensated by very low taxes that permit a final retail price not so elevated in comparison with the rest of Europe.

3.1.1.2 Marginal Cost

This range of prices is interesting and will be of great use in this study, nevertheless we need to pay attention to the fact that they represent the average cost, and this is not the cost we are the most interested in. Indeed, we are evaluating electricity price in order to further determine the amount of money a customer could save on his bill by installing solar panels, and this sum of money will not depend on the average cost but on the marginal cost. To explain this let us take a concrete example and examine in Table 3.1 the details of an electric bill in Spain, extracted from Sancha Gonzalo's work [2].

Table 3.1 Detail of an electric bill in Spain (2012)

Energy	(1) energy price	0.06746 €/kWh
Energetic systems and policies	(2) access toll to energy	0.06367 €/kWh
	(3) toll relative to the contracted power	16.63313 €/kW/year
	(4) capacity cost	0.01119 €/kWh
	(5) commercialization cost	4€/kW/year
	(6) cost of hiring the meter box	0.54 €/year
Taxes	(7) electricity special taxes	4.864%*1.05113*[(1)+(2)+(4)+(5)]
	(8) VAT	18%*[(1)+...+(7)]

	variable costs
	fixed costs

Average cost	0.22014 €/kWh
Marginal cost	0.17740437 €/kWh

Source El Sistema Eléctrico Español (IV). Sancha Gonzalo [2] and own elaboration

The final price of electricity can be divided in three main components. The first one represents the direct cost of energy and is proportional to the number of kWh consumed during the year. Contrary to what we could think, this is just 30% of the average cost. Then the second component gathers costs related to the grid maintenance and to the connection of the household to the grid. In this section there are variable costs that depend on the annual consumption and fixed costs that will not change in function of the consumption but are proportional to the contracted power. This is where the difference between average cost and marginal cost appears. Indeed, the marginal cost corresponds to the additional price if one more kWh is consumed, consequently to calculate it the fixed costs are not taken into account. Lastly, the third component includes the taxes. In Spain, two taxes are applied on electricity, the VAT like on any other product and the special electricity taxes. The second one is applied only on some components (see details in Table 3.1) whereas the first one on every of them. These taxes are taking into account in both the calculation of the average and the marginal cost.

Finally, an average cost of 0.22 €/kWh and a marginal cost of 0.18 €/kWh are obtained, which makes a difference of 19% between both prices. This difference is significant and will play a key role when it comes to evaluate the profitability of a PV system. Table 3.2 gives us the value of the average and marginal costs in three other Mediterranean countries. It turns out that the difference between both costs is quite similar in every case.

Moreover, it is worth reminding why marginal costs are more relevant for the present study than average costs. We are estimating the money a householder could save on his electrical bill by buying and installing a PV system. As this system will not cover all his energy demand, the client will stay connected to the grid and will keep consuming from it when needed. This is why the annual fixed costs on his bill will not change, he will just save the marginal cost corresponding to the electricity generated by the solar panels instead of taking it from the grid.

3.1.1.3   2016 Prices in Spain

Sancha Gonzalo's work [2] is very interesting as it explains in detail the different components that are taking part in the final electrical price. However, the paper is from 2012, and for our study more recent prices are needed. This is why we looked

Table 3.2 Difference between average and marginal cost, price in €/kWh

	Average cost	Marginal cost	Difference (%)
Spain	0.22	0.18	19.4
France	0.15	0.12	19.0
Portugal	0.21	0.17	16.9
Italy	0.26	0.22	17.5

Source El Sistema Eléctrico Español (IV). Sancha Gonzalo [2] and own elaboration

Table 3.3 Electricity 2016 retail prices offered by different companies, prices expressed in €/kWh

	Energy term	Electricity special taxes	VAT	Retail price
Goiener	0.137	0.007004394	0.02592079	0.170
Gesternova	0.118	0.006032982	0.02232594	0.146
Gas natural fenosa	0.135	0.00690214	0.02554239	0.167
EDP	0.12	0.006135236	0.02270434	0.149
Viesgo	0.17	0.008691584	0.03216449	0.211
Iberdrola	0.13	0.006646505	0.02459637	0.161
SOM energía	0.124	0.006339743	0.02346115	0.154
Holaluz	0.123	0.006288616	0.02327195	0.153
Endesa	0.14	0.007157775	0.0264884	0.174
Pepeenergy	0.112	0.00572622	0.02119072	0.139
Average	0.1309	0.006692519	0.02476665	0.162

Source Energía info [3] and own elaboration

for public prices directly provided by the distribution companies themselves. Indeed, these prices are always actualized, whereas prices quoted in paper are most of the time obsolete. We will first deal with the Spanish case, where it exists many private energy distributors in addition to the three original leaders in the market: Endesa, Iberdrola and Gas Natural. Table 3.3 gathers some of these distributors and the electricity price they respectively offer.

The price published by each company is not the retail price (or end user price), but corresponds to the energy term. To obtain the retail price we have to add taxes. As previously explained, there are two types of taxes we have to take into account: the electricity special taxes and the VAT which is 18% in Spain. In the next part of this study, the average retail price that figures in Table 3.3 will be used as the reference price for the marginal cost of electricity in Spain.

We can note that the price band offer is quite wide. Indeed, the lowest price is offered by Pepeenergy at 0.139 €/kWh for the end price, while the highest is from Viesgo at 0.211 €/kWh. We could wonder how this last one can survive with such a high price in comparison with the competition. Actually, it is because this tariff is just one of those offered by the company which also proposes tariffs with time discrimination. And without doubt, its policy is to encourage clients to subscribe one of this new generation tariffs, this is why they put the classical fixed tariff quite high. In the next paragraph these new kinds of contract will be presented.

3.1.1.4   Tariffs with Time Discrimination

The apparition of digital meter box (or smart meters) opened many new opportunities for energy distribution companies. One of their objectives is to reduce the two daily peaks in the domestic demand (see Fig. 3.3). For this, they created tariffs with

Fig. 3.3 Profile of domestic electric demand in function of the tariff chosen. *Source* Disposición 3069 del BOE número 68 de 2016 and own elaboration

time discrimination, which means that the electricity price is not the same in function of the hour at which it is consumed. We will detail here the case of Spain where it exists two different kinds of contract with time discrimination for particulars.

The first one is called the 2.0 DHA tariff and considers two distinct pricing periods. During peak hours, which approximately correspond to the day, the electricity is more expensive than during off-peak hours (at night). Table 3.4 shows the exact time periods that are slightly different between summer and winter.

As we previously did for the classical tariff, we are going to consider the average price calculated from the offers made by the different companies existing on the market. Table 3.5 displays the major companies' offers and the average price obtained. During peak hours clients that have subscribed to this contract will have to pay 0.19 € for one kWh, which is more than with a normal tariff (0.162 €/kWh), but during off-peak hours they can take advantage of a much cheaper price: 0.09 €/kWh.

The second tariff is the 2.0 DHS, it is the same system but with three time periods instead of two. Not all the companies propose this contract to its clients, this

Table 3.4 Time periods for 2.0 DHA and DHS tariffs

	Peak hours	Off-peak hours	Super off-peak hours
2.0 DHA	12 am–10 pm in winter 1 pm–11 pm in summer	10 pm–12 am in winter 11 pm–1 pm in summer	–
2.0 DHS	1 pm–11 pm	11 pm–1 am and 7 am–11 am	1 am–7 am

Source Own elaboration

Table 3.5 Peak and off-peak hours prices offered by different companies (marginal cost, taxes included)

		Marginal cost (€/kWh)
Goiener	Peak hours	0.187
	Off-peak hours	0.097
Gesternova	Peak hours	0.173
	Off-peak hours	0.078
Gas natural fenosa	Peak hours	0.202
	Off-peak hours	0.110
EDP	Peak hours	0.176
	Off-peak hours	0.074
Viesgo	Peak hours	0.224
	Off-peak hours	0.114
Iberdrola	Peak hours	0.203
	Off-peak hours	0.087
SOM energía	Peak hours	0.179
	Off-peak hours	0.084
Holaluz	Peak hours	0.192
	Off-peak hours	0.098
Average	Peak hours	0.192
	Off-peak hours	0.093

Source Energía info [3] and own elaboration

is why the average price for each period depends only on two offers. We can see that during peak hours electricity is even more expensive than with the 2.0 DHA, but it is balanced with the 0.08 €/kWh price available between 1 am and 7 am (Table 3.6).

The 2.0 DHA or DHS tariffs are profitable only for clients that have a high electrical consumption during the night. It can be the case for households with an electric vehicle or for households willing to change their consumption habit. For instance, domestic appliances like washing machine or dishwasher can be programmed to run at night. On Fig. 3.3, you can see the strong difference between the charge profile for clients with or without time discrimination. However, in some cases, habit cannot be changed so drastically, for example we can think about lights and heating systems that are needed during the afternoon.

Table 3.6 Tariffs 2.0 DHS offered by two different companies

	Peak hours	Off-peak hours	Super off-peak hours
Gas natural fenosa	0.163	0.089	0.068
Iberdrola	0.161	0.083	0.056
Average	0.162	0.086	0.062

Source Energía info [3] and own elaboration

3.1.1.5   The Fixed Costs of an Electrical Bill

We saw in Sect. 3.1.1.2 that the fixed costs of electricity depend on the contracted power chosen. These costs represent an important part on the total electric price. In the example given by Sancha Gonzalo [2] they represent, for Spain, 20% of the total amount of the bill. Consequently, someone wishing to make savings on his energy facture needs not to focus only on the variable costs. Unfortunately, buying and installing a PV system do not enable to reduce these costs. This paragraph aims at explaining why.

The profile of the daily domestic demand in electricity contains two peaks, one around midday and the other around 8 or 9 pm (see red curve of Fig. 3.2). The hired power needs to be sufficiently high so that the fuses do not trip during these peaks. Installing a PV system can help to pass the first peak of the day as it occurs during sunned hours. However, the second peak happens at night, when the PV system cannot run, consequently the hired power must be calculated without taking in account the presence of solar modules.

Determining the hired power necessary in a home is quite simple. You must do the list of all important electrical appliances with their respective power. Then, you have to sum the powers of the different devices and to add 1 kW to take in account all the small electric equipment (lights, hair-dryer, laptop…). Finally the hired power required corresponds to this amount divided by 2 which is the simultaneity factor (all the devices will not run at the same time). Table 3.7 shows an example of this calculation for a typical Spanish dwelling. It turns out that a hired power of 5.6 kW is necessary. Typically companies offer a contract with 5.5 kW of hired power, this would be the right choice in this case whether the house disposes or not of a solar installation. Sometimes, the variable cost of one kWh of electricity depends on the hired power subscribed by the client. This is because companies do not offer the same price to big consumers as to particulars. All electric prices that appear in this paper are considered relatively to a 5.5 kW hired power, which is the normal amount for the kind of houses we are interested in.

Table 3.7 Details for calculation of an average contracted power

	Device's power (W)
Fridge	300
Micro-waves	1200
Washing machine	2000
Dishwasher	2000
Oven	1500
Electric stoves	1000
TV	200
Heating or cooling system	2000
Divers	1000
Total	11,200
Hired power required	5600

Source Own elaboration

3.1.2 Solar Irradiation

Solar irradiation is one of the major parameter to take into account when it comes to design a PV system. Indeed, the amount of energy produced by a similar system will greatly vary in function of its geographic position. For instance, the Northern countries will receive few irradiation during the year because of their latitude and the cloudier weather, consequently the solar production there will be much lower than in a Mediterranean country.

Before starting, it is worth reminding the slight difference between solar irradiance and irradiation, both terms which are often mixed up. The first one corresponds to the power per unit area received from the Sun in the form of electromagnetic radiation, it is expressed in W/m^2. By integrating over time this variable, we obtain solar irradiation, which is therefore homogeneous to an energy (Wh/m^2). In this paper, we will always talk about this former one. The JRC European commission provides us a fantastic tool that allows to directly obtain the local irradiation of a given geographic position. This tool will be used in the present study to determine the amount of electricity a solar panel can produce in each European country and to plot daily solar curves.

Table 3.8 gives us a few examples of the irradiation received during one year in different places of Europe.

However, we will see later that with an irradiation of 1000 kWh/m^2 and one square meter of solar modules, we cannot produce 1000 kWh of electricity. There are losses due to the system and the real amount of electricity generated is fairly inferior.

One important thing to know when it comes to install solar panels is the best orientation of the panels in order to optimize the production. This optimum orientation depends on the latitude, lower is the latitude, smaller the orientation angle (between the ground and the panel, see Fig. 3.4) must be. Additionally, due to the rotation of the Earth, the optimum slope change from a season to another, that's why it exists some modules that are mounted on a mobile structure which adapts its orientation in function of the hour and the day. But in our study we will consider only panels with a fixed slope, consequently the orientation chosen has to be an

Table 3.8 Examples of local irradiation in Europe		Global irradiation (kWh/m^2)
	Paris	1420
	Madrid	2030
	London	1280
	Rome	2040
	Berlin	1260
	Cyprus	2180
	Copenhagen	1210

Source JRC European commission [4] and own elaboration

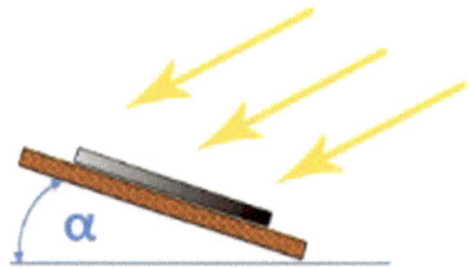

**α is the angle to optimize to obtain the best
efficiency for the panels**

Fig. 3.4 Drawing of the module's slope. *Source* Own elaboration

average of the optimum slope for each month. The tool provided by JRC makes the calculation for us, it turns out that within Europe the optimum angle doesn't vary much between a country and another. Actually it is evaluated at 41° for Scandinavian countries and at 33° for the Mediterranean ones.

This tool also allows us to obtain the curves of daily radiation that we will use in further sections to determine the capacity of our system. Figure 3.5 displays these curves for Madrid area, for the months of January and July, considering a 1 kWp installation. Without surprise, the radiation is way stronger during summer than during winter, but what it is interesting to note is the change in the hour of peak production. This peak will later be compared to the peaks in household electricity demand, which unfortunately do not occur at the same hour.

By calculating the area of these both curves we realize that the global irradiation in summer is two times higher than in winter. This simple observation highlights one of the main problems of photovoltaic: the periods of major production do not fit

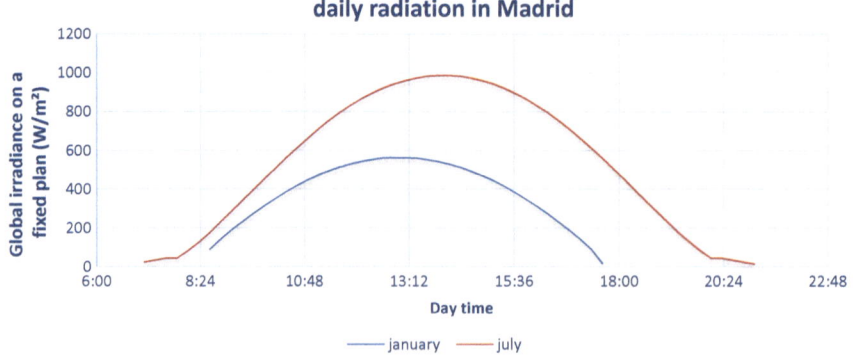

Fig. 3.5 Daily irradiance in Madrid. *Source* JRC European commission [4] and own elaboration

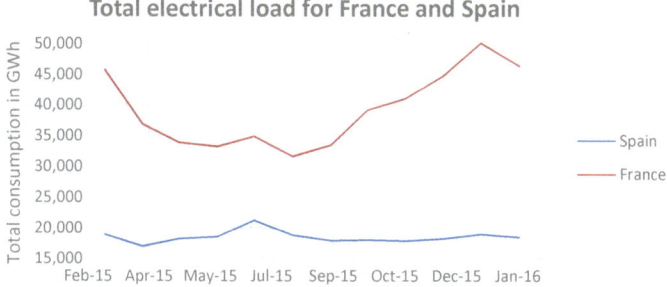

Fig. 3.6 Electric load perfil for Spain and for France. *Sources* REE [5], RTE [6] and own elaboration

with the periods of major demand. In Spain the total load is superior in summer than in winter because of the high need in air conditioning. But in many other European countries the contrary happens, the consumption of heating systems causes a higher demand in winter than in summer (See Fig. 3.6 with example of France and Spain).

3.1.3 Components of the System

In this section, we will describe the PV system on which is based the whole model. Technical characteristics and explanations about their functioning will be given, however the fabrication process will be detailed only in Chap. 5. This choice was made because the understanding of this process is useful mostly for the elaboration of the carbon footprint but is not essential to this section which is focused on the operational function of the system.

From now on, we will often deal with the installed capacity of the PV system, so a brief definition is necessary. It is expressed in Watt-peak (Wp) and represents the theoretical capacity of the whole installation, for instance a 1 kWp installation will generate 1 kWh of electricity under standard conditions (an irradiation of 1 kW/m^2 during one hour, with a slope of 35° and the temperature of the cells being 25 °C). In the literature it is also referred as the nominal power.

3.1.3.1 Modules

The PV modules constitute the core of the system. On the market it exists a lot of different models but they all have quite similar technical features. For our study we will consider modules with typical characteristics that are described in Table 3.9.

One module of this kind delivers a power of 250 W, the number of modules in the system is determined in function of the installed capacity required. For instance, for a 2 kWp installed capacity 8 modules are needed. The modules are made of

Table 3.9 Technical characteristics of the solar panel

Power rating	250 W
Efficiency	15%
Number of cells	60
Type of cells	Polycrystalline silicon
Surface area	1.6 m^2
Weight	20 kg
Life span	25 years
Certified to 90% efficiency	10 years
Certified to 80% efficiency	25 years

Source Own elaboration

polycrystalline silicon cells which is the technology normally used for domestic solar panels. Compared to the monocrystalline cells it has the advantage to be costless, to waste less silicon during the formation process and to be less sensitive to dirt while functioning. The disadvantage is that polycrystalline cells tend to have a lower efficiency than monocrystalline ones which could reach a 20% efficiency. However, for a domestic use, this slightly higher efficiency does not compensate the higher costs and this is why polycrystalline cells are chosen.

The lifespan of the system is 25 years, consequently buying and installing a PV system must always be in the framework of a long term project and in the further sections of this study the economic analysis will be done over 25 years. Moreover, modules don't lose much efficiency over their life as the constructors guarantee a functioning at 90% of its capacity after 10 years and at 80% after 25 years.

3.1.3.2 System of Control

The profiles of the household demand curve and of the electricity generation from the PV system are totally different (see further Fig. 3.10). It means that there are periods in which the production cannot satisfy the load, typically during night hours, in this case the electricity needed is bought to the grid as it would be done without any PV system. And there are other moments in which the demand goes beyond the production, typically during midday as the consumption is low and the sun very powerful, then the excess of energy needs to be supplied somewhere. A solution could be to burn this energy in a random resistance. But this appears as a waste of energy and as a useless wear of the modules which do not have a limitless lifespan. Consequently we chose to implement a system of control in our PV installation.

The role of this system of control will be to light on or to light off some cells of the panel in function of the instantaneous demand. The algorithm ruling the system is displayed in Fig. 3.7. For this algorithm, the modules are numbered from 1 to n (n being the total number of modules in the installation), and the variable responsible for this numeration is called x. The increment of x at each loop is made

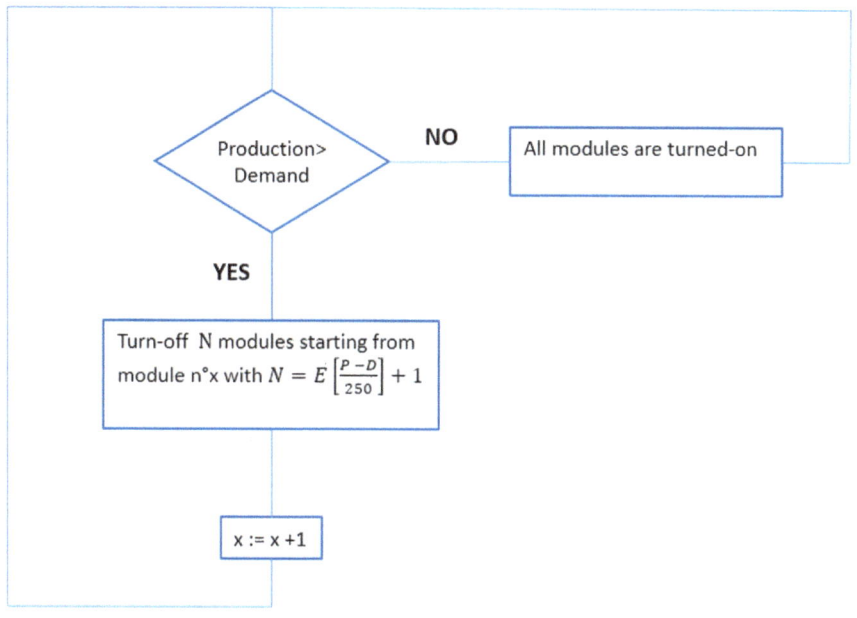

Fig. 3.7 Algorithm of the system of control. *Source* Own elaboration

to assure a homogeneous utilization of each module (if the same module is turned-off each time the production becomes higher than the demand, it would wear out less than the others).

3.1.3.3 Inverter

The inverter is an essential component of a solar installation, it permits to convert the DC power generated by the modules into AC power usable for the domestic equipment. It represents a significant part of the total cost, around 15–25%, and it has a life expectancy much shorter than the modules, estimated at 10 years. So, in our model we have to take into account the buying of two new inverters during the 25 years of functioning.

For PV systems which do not include a battery, three main types of inverter exist on the market. Micro inverters are the less common and the more expensive. Their advantage is that one box is installed on each module and it converts right at the panel the energy produced into AC power. It means that if one module is performing at a lower level than the others it will not jeopardized the whole installation. This can happen in case of a failure but also in case of a shaded installation.

The two other types, which take a significant part of the market (approximately 90% for both of them), are the central inverters and the string inverters. Figures 3.8 and 3.9 respectively show the way they operate. Central inverters require fewer

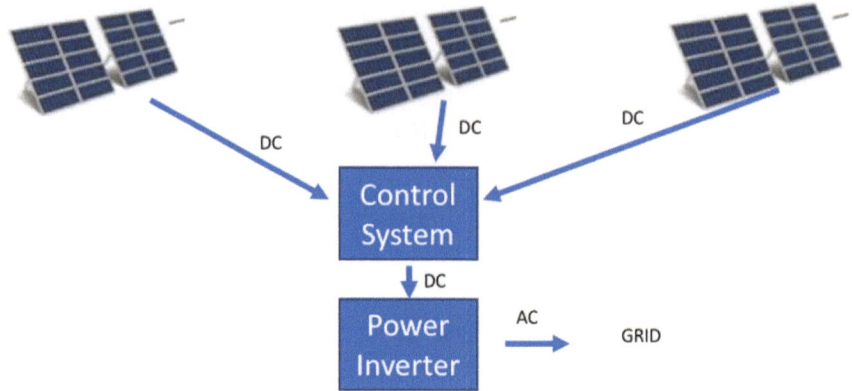

Fig. 3.8 Technical scheme of a central inverter. *Source* Own elaboration based on [7]

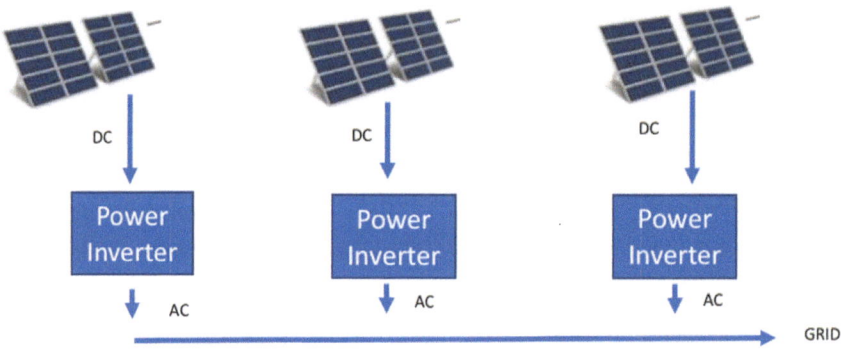

Fig. 3.9 Technical scheme of a string inverter. *Source* Own elaboration based on [7]

component connections and carry the power under its DC form towards a central box which converts it into AC power. In the other case, there are multiple smaller inverters for several strings, so the DC power from a few strings runs directly into a string inverter rather than in a combiner box and is then converted to AC. Both systems have similar global costs, so the choice cannot be made in function of that. It turns out that central inverters are more adapted to large systems where production is consistent across array. While string inverters, with a lower maintenance cost and a better modularity in case of a non-constant production between the different panels, are more likely to be used in domestic or small installations. This is why, in our model, it is this last kind of inverters which is chosen.

3.1.4 Legalization Procedures

Buying a PV system is not like buying any other domestic equipment, it requires some administrative procedures that can be sometimes long and hard. In this section, we will display the procedures necessary to install solar panels for self-consumption in Spain. These processes are regulated by the three following norms: RD 900/2015, RD 1699/2011, RD 1955/2000. They can be summed up in three main points. The first thing to do is to ask to the distribution company for the installation of a new connection point to the grid. If the PV system has a capacity inferior to 10 kWp and no energy will be transferred back to the grid, the client is exempted from taxes on this installation and the company cannot refuse to do it.

The second point is the obligation to include in the whole system a measurement device that quantify the net energy generated by the solar panels. In the next paragraph, the cost of this device will not appear directly, it is contained in the installation package. Finally, the last thing to do is to register the installation in the administrative list of electricity self-consumption ("registro administrativo de autoconsumo de energía eléctrica") which is handled by the "Dirección General de Política Energética y Minas del Ministerio de Industria, Energía y Turismo". This inscription can be done in the city hall and is not charged.

As a result, for a PV system with a capacity under 10 kWp which is destined only to self-consumption, there is no direct cost of legalization. However, it is a tedious process that can have two consequences. The first one is that in Spain many particulars have not declared their solar installation and therefore use it in an illegal way. The second one is that someone wishing to invest in PV energy can be discouraged by these administrative procedures. This is why many actors of the solar sector are asking for a simplification of these norms which represent more a brake to the market's development than anything else.

3.1.5 Net Investment: Costs of Equipment

For this study, we will consider the price of a PV system similar in every European country. Indeed, within Europe there is no taxes or legal problems for carrying the system from one country to another if needed. However, evaluate the different costs is quite difficult because of the wide range of prices that exist on the market. On the web, the majority of the sellers offers all-included kits with prices that do not reflect the decline in PV technology costs during the last decades.

Specialists of the field speak about PV prices in terms of €/Wp, that is to say the price of every component depends on the installed capacity which is chosen for the system. For each element a price band is considered, as it would be too risky to simply consider a fixed price, these ranges of prices are featured in Table 3.10. Consequently, we will further take in account for PV system's price two different scenarios. A first one with the highest prices of the band that could correspond to a random client that does not want to spend too much time looking for the best market prices and that would buy on the internet after a rapid benchmarking. Then the second scenario stands for a type of client more informed and with more experience in the field that could reach lower prices by buying the components separately or doing a part of the installation by himself.

Considering the bracket prices given in Table 3.10 we are going to determine the respective price functions of each scenario. The objective is to express the total price of the system as an affine function depending on the installed capacity. It will further allow us to choose the capacity that best optimizes the profitability of the installation.

In addition to the costs of the components we have previously presented (modules, system of control and inverter), we have to take in consideration two more costs: the structure that permits to fix the panels either on a roof or on the ground and the installation that generally needs to be done by a professional (Table 3.11).

Table 3.10 Range of prices for the different components of the PV

	Price in €/Wp
PV modules	0.5–1
Structure	0.07–0.1
Inverter	0.2–0.3
Installation	0.05–0.5
System of control	0.1–0.2
Total	0.92–2

Source Bachelor's thesis "Viabilidad del almacenamiento energético en instalaciones fotovoltaicas de autoconsumo", author Guisado, J. M. supervisor Lillo, I., University of Seville (2016) [8] and own elaboration

Table 3.11 Function prices for the whole system (in €)

	High prices scenario	Low prices scenario
Module 250 W	250	125
Inverter	800	600
System of control	500	300
Installation	800	500
Structure	400	300
Total	2500 + 250 * N	1700 + 125 * N

Source Bachelor's thesis "Viabilidad del almacenamiento energético en instalaciones fotovoltaicas de autoconsumo", author Guisado, J. M. supervisor Lillo, I., University of Seville (2016) [8] and own elaboration

We finally obtain the following functions:

$$\text{High prices scenario:} \boldsymbol{Total\ price}\ (\text{€}) = \boldsymbol{2500 + 250\ 2} * N$$
$$\text{Low prices scenario:} \boldsymbol{Total\ price}\ (\text{€}) = \boldsymbol{1700 + 125\ 2} * N$$

where N stands for the number of modules.

For instance for a 3 kWp installation, 12 modules (of 250 Wp each) are needed and the total cost is in one case 5500 € and in the other 3200 €. Transforming this price in €/Wp it gives respectively a total cost of 1.83 €/Wp and 1.06 €/Wp, which enters in the price range provided by Table 3.10. However you can note that in the HP scenario if we consider a system with a low installed capacity we will get a price out of this range. For example for a 1.5 kWp installation, the total cost is 4000 € which corresponds to 2.6 €/Wp. But it makes sense to consider this price. Indeed, this scenario stands for a type of client that would buy a turnkey project, it means that a company would come to the house for the installation and probably come back every year for the maintenance. If the company charges a total price in the range 0.92–2 €/Wp they cannot amortize the travel, this is why they overcharge the modules' price so as to cover their costs.

3.2 Optimization of the Capacity

This section aims at determining the installed capacity that optimizes the economic profitability of the PV system. The optimum capacity is the one that corresponds to the best compromise between a high capacity which can fulfill all the demand but is expensive and produces many losses, and a low capacity that is cheaper but cannot meet all the needs. To find this compromise, we will design a model which calculates the economic profitability in each case, this is the object of Sect. 3.2.2. Before this, in Sect. 3.2.1 we present a method that directly estimates the capacity required in function of the electrical needs and the local irradiation. This method does not take into account any parameter of optimization between consumption and losses, this is why we will see that its results are not totally satisfying.

3.2.1 Direct Calculation Method

This method is a simple calculation (Eq. 3.1) which gives us an estimation of the capacity we should install to fulfill the demand in electricity in function of the geographical position. It comes from the work of Pablo Eguia, Esther Torres and Javier García in their project "gedisper" [9].

$$C_p = \frac{Annual\ charge * Panel's\ capacity}{efficiency * irradiation * panel's\ surface} \tag{3.1}$$

- The annual charge is the average electric consumption for a household during one year, it is expressed in kWh and we used data from the World Bank data [10] that offers an evaluation of this consumption for almost every country of the world.
- Panel's capacity and surface correspond to the panel we chose in the anterior section, they are respectively 250 W and 1.6 m^2.
- The efficiency refers to the portion of energy received from the sun the system can convert into electricity. We used the value 15% that is standardized for solar cells.
- The irradiation expressed here in W/m^2 is the total amount of energy that will be received by one square meter of panel during one year. The data comes from JRC Europa as we explained it in Sect. 3.1.2.

Table 3.12 displays results for a few European countries. It turns out that, according to this model, in Mediterranean countries like Spain, Italy o Malta an installed capacity between 1 and 2 kWp would be enough to fulfill the household's demand. However, in Northern countries there is a lower irradiation and a higher consumption because the needs in heating and lighting are very important in winter as the temperatures are very low and the nights very large, this is why a 7–8 kWp capacity would be necessary in these areas.

But the problem with this basic model is that it doesn't take into account that all the electricity produced by the solar system cannot be directly used to fulfill the house electrical needs. Actually, as we explained in Sect. 3.1.2, the load curve and

Table 3.12 Required capacity according to Gedisper's method

	Peak capacity (W)
France	3801.7
Spain	1972.1
Germany	2626.1
Finland	7575.8
Malta	1705.4
Polonia	1570.8
Portugal	1769.8
UK	3276.2
Sweden	7523.1
Italy	1373.6
Greece	2050.1
Austria	4130.7

Source Own elaboration

the solar production curve are not similar at all, the PV system has its peak production at midday which is an off-peak time for domestic electrical charge. This is why, with installed capacities such as these featured in Table 3.12, a great part of the electricity produced would be lost. Consequently, we designed another model that considers these losses and try to find the capacity which optimizes the profitability of the system. And with this second model we will see that capacities presented in this section are way too high to be profitable under our hypothesis.

The model presented in the project "gedisper" would be interesting if energy storage were considered. Indeed, as in this case the PV capacity is designed to fulfill all the household's demand, with the introduction of a battery in the facility, the client could hope to become self-sufficient in electricity.

3.2.2 Optimization Model for Installed Capacity

In a first phase, we are going to elaborate the model for the Spanish case, the international comparison will be done in Chap. 4.

3.2.2.1 Estimation of Losses in Function of the Installed Capacity

Before starting, we remind that what is called losses in this study correspond to the amount of energy that is produced by the solar system but that cannot be consumed because the demand is inferior to the production. One way to take advantage of these losses would be to use energy storage or to sell it to the grid, but this is not the purpose of this project.

To evaluate these losses, the solar production curve is placed on top of the load curve and the intersection area between both curves is calculated (see Fig. 3.10).

Both the electrical demand and the solar production vary over the year. Indeed, in summer there are more sunned hours and the sun is more powerful so the production is way higher than in winter. Likewise the load curve in winter is different from the summer one as the needs are different (in summer the curve tends to be smoother than in winter). Therefore, we draw the curves as in Fig. 3.10 for each month of the year. In the next paragraphs we are going to explain how we obtained both type of curves.

• Load curves

These curves were obtained thanks to data collected thanks to smart meters from a real typical client. Smart meters are currently being installed in all Europe, they permit to gather a lof of information on consumers' habits, and therefore are the base of many studies aiming at reducing both the consumption of energy and the price of the bills. In our case, the data used are published in the "Disposición 3069 del BOE número 68 de 2016" [11], a document edited by the Spanish minister of

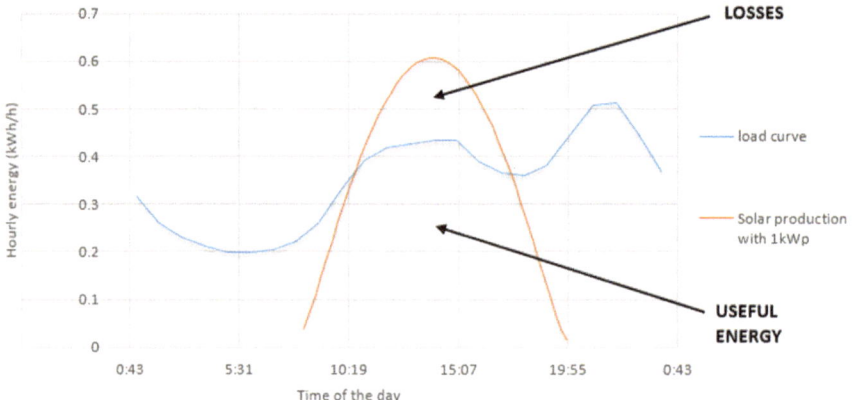

Fig. 3.10 Electric load compared to solar production in March in Madrid. *Sources* Disposición 3069 del BOE número 68 de 2016 [11] and Joint Research Centre, European commission [4] and own elaboration

presidency and territorial administrations. This file provides for each hour of the year its contribution to the annual consumption, in other words each hour of the year corresponds to a certain percentage of the total consumption. Then, to get the daily curves we have just multiplied each percentage (also called weight) by the annual consumption. In Spain the average annual electrical consumption for a household is 3250 kWh [12].

Figure 3.11 shows the superposition of the 365 curves of the year, we can clearly see that at any moment of the year there are two peak hours, and that the curves all have more or less the same profile. For our model, we picked one representative curve for each month so as to put it on top of the solar energy production curve and obtain 12 graphs similar to the one in Fig. 3.10.

- Solar production energy curves

To create these curves we relied on an online tool provided by the "JRC European commission" [4] that was presented in Sect. 3.1.2. Thanks to this tool we obtained for each month a set of data like showed below in Table 3.13.

Then to avoid unity problem we used a system of weight like we did for the load curves. The weight for each fifteen minutes corresponds to the global irradiance at this time divided by the total of the irradiance for the day (14,162 W/m^2 in the example). Finally the curve we want corresponds to the last column of Table 3.13. It is obtained by multiplying the weight by 4 (because it is quarter of an hour energy whereas for the load curves we have hourly energy) and by 2.83 which is the daily electricity produced by a 1 kWp PV system in Madrid in January. We did this process for capacity between 0.5 and 2.5 kWp and for each month (Fig. 3.12 represents the results for January) so as to obtain a percentage of losses in each case. Theses percentage are gathered in Table 3.14.

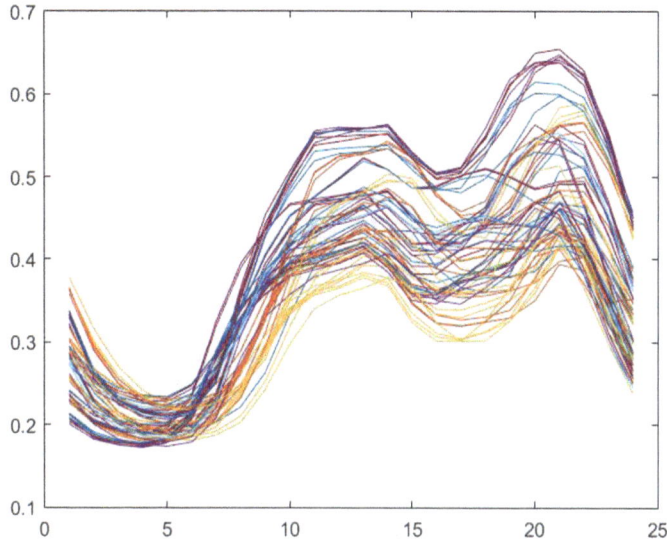

Fig. 3.11 Superposition of the 365 daily load profiles over the year. *Source* Daniel Lugo Laguna's master thesis [13]

Table 3.13 Data used to elaborate the daily profile of solar production, for January in Madrid

	Global irradiance (W/m^2)	Weight (without unity)	Curve for 1 kWp (kWh)
8:37	92	0.00649626	0.07353764
8:52	144	0.01016806	0.11510239
9:07	189	0.01334557	0.15107188
9:22	232	0.01638187	0.18544273
⋮	⋮	⋮	⋮
16:52	189	0.01334557	0.15107188
17:07	144	0.01016806	0.11510239
17:22	92	0.00649626	0.07353764
17:37	50	0.00353057	0.03996611
Total for the day	14162		

Source Joint Research Centre, European commission [4] and own elaboration

In Table 3.14 we can note that of course higher is the installed capacity higher are the losses, but what is more interesting to note is that for a given capacity the percentage of losses is quite stable for months between March and November. For instance, a 2 kWp installation will lose half of its production during three quarters of the year. The winter months make exceptions, as there is very few sunshine during this period the losses are less important.

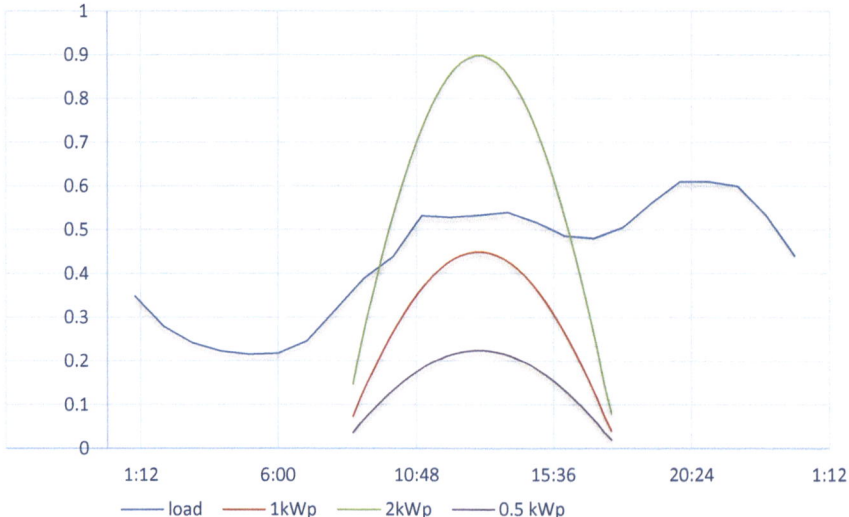

Fig. 3.12 Comparison of the load profile and the production with different installed capacity, in January, in Madrid. *Sources* Disposición 3069 del BOE número 68 de 2016 [11] and Joint Research Centre, European commission [4] and own elaboration

Table 3.14 Losses in percentage of the total solar production, in Madrid

Losses					
	0.5 kWp (%)	1 kWp (%)	1.5 kWp (%)	2 kWp (%)	2.5 kWp (%)
Jan	0.00	0.00	9.12	27.35	40.17
Feb	0.00	2.17	26.86	42.79	53.55
Mar	0.00	18.04	42.14	54.91	63.22
Apr	0.00	18.29	41.63	55.00	63.03
May	0.00	18.41	40.90	53.55	62.15
Jun	0.00	20.22	41.60	54.60	62.40
Jul	0.00	20.90	43.43	56.03	63.92
Aug	0.00	18.98	41.35	54.58	62.86
Sep	0.00	18.58	41.10	53.79	62.31
Oct	0.00	18.47	41.23	53.99	62.39
Nov	0.00	11.69	35.57	49.21	59.06
Dec	0.00	0.00	7.77	25.11	37.58

Source Own elaboration

Considering these losses we will now differentiate the energy produced from the energy that is truly consumed (called useful energy). The produced energy is a linear function of the installed capacity, whereas the useful energy is an increasing function but which has an upper limit. It means that once a threshold is reached,

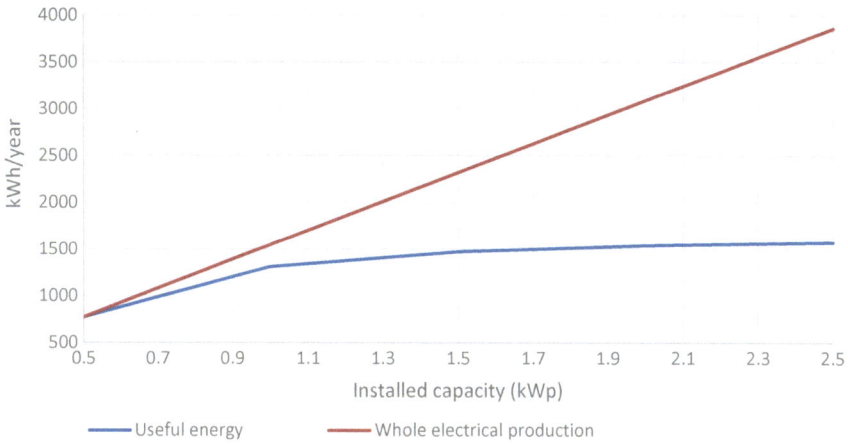

Fig. 3.13 Difference between energy produced and useful energy for each installed capacity. *Source* Own elaboration

even if you choose a superior capacity you will not take advantage of more consumed electricity. Both curves appear on Fig. 3.13.

We can see that the amount of energy that can be consumed is not very different with a 2.5 kWp capacity than with a 1.5 kWp. It means that a 2.5 kWp installation is more expensive than a smaller one but does not necessarily make the householder save much more money on his electrical bills. Therefore an economic analysis needs to be done to determine which capacity will be more profitable under these conditions. This is the purpose of the model presented in the next paragraph.

3.2.2.2 Financial Indicators

To determine which installed capacity is the more profitable the net present value (NPV) and the internal rate of return (IRR) of the installation are calculated in each case. The study is done over 25 years which corresponds to the life time of the solar panels. In the calculation of the NPV, the WACC (weighted average cost of capital) is considered worth 0%, which is the same as considering that the investment made by the particular has no cost of opportunity. This choice was made because the amount of money invested is quite low, thus, the client would not necessitate a credit nor would change his way of living in case of realizing the investment. Additionally, many governments are offering public aids to help financing these investments. It generally consists in loans without interest.

The first thing to do is to determine which the income and the costs related to the investment are. The following paragraphs will detail the different hypothesis that have been done to do so.

- The costs

There are two main types of costs associated to a PV system: the initial investment and the cost of maintenance and operations (that were mentioned as *Opex* in the literature review).

A PV installation requires very few maintenance. The main task consists in cleaning the modules because the accumulated dirt on top of them can reduce the system's efficiency. Consequently the maintenance costs are very low, they are generally expressed in function of the initial cost of the equipment. According to what Bhandari and Stadler [14] did, we will consider them at 1% of the initial cost (as another example, Hernandez-Moro and Martinez-Duart (2012) [15] consider them at 1.5%)

At the initiation of the project the net investment corresponds to the total price of the equipment that we have detailed in Sect. 3.1.5 and separated in two scenarios: high PV prices (HP) and low PV prices (LP). Additionally we have to take into account another cost linked to the inverter. This component has a lifespan of only 10 years, this is why it needs to be replaced at year 10 and year 20 of the project.

- The income

In this project the annual income actually corresponds to savings as it depends on the amount of electricity the client can use from his solar installation and therefore does not need to buy to the grid. This amount of electricity, that we called useful energy, depends on the installed capacity and was determined in the anterior section. Table 3.15 reminds the main values.

Then to obtain the annual income this amount of energy is multiplied by the price of one kWh. As we explained it in Sect. 3.1.1.2, we consider here the marginal cost of electricity which is currently 0.162 €/kWh for Spain. You will find in Table 3.15 the corresponding income for each installed capacity. It is interesting to note that with a 2.5 kWp capacity, the annual income is only 6 € higher than with a 2 kWp one, whereas it requires the buying of two 250 W modules more.

However these are the incomes just for the system's first year of functioning as we take into account a slight diminution of its productivity over time. We relied on the constructor's guaranty which promises a 90% efficiency after 10 years and an 80% one after 25 years.

Finally, in this first scenario, the hypothesis of a constant electricity price is made. Later we will consider two other scenarios and it will turns out that the model is very sensible to the variation of the electricity cost.

Table 3.15 Amount of money saved on the electric bill in function of the installed capacity

	0.5 kWp	1 kWp	1.5 kWp	2 kWp	2.5 kWp
Useful energy (kWh)	772	1309	1470	1537	1571
Savings on electrical bill (€)	125	212	238	249	255

Source Own elaboration

Table 3.16 Hypothesis for a first application of the model in Madrid

Electricity	Marginal cost = 0.162 €/kWh, constant over years
Useful energy	In function of the capacity (see Table 3.15), linearly decreasing until 90% after 10 years and until 80% after 25 years
Cost of equipment	LP scenario, 1700 + 125 * N, N being the number of modules
Inverter	Lifetime = 10 years, price = 600 €
Maintenance	1% of the system's initial cost (annually), constant over time

Source Own elaboration

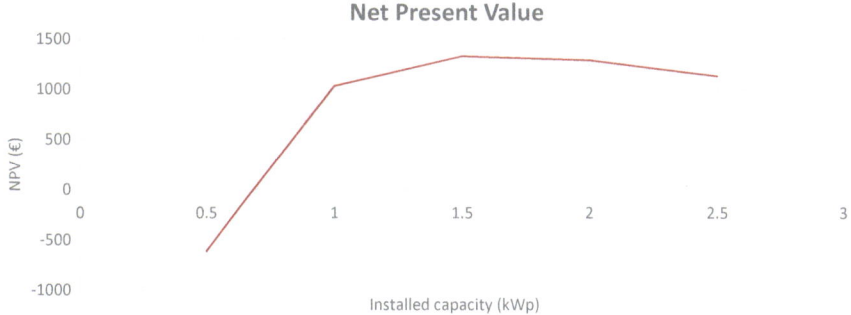

Fig. 3.14 NPV of the PV system in function of the installed capacity for Madrid area. *Source* Own elaboration

- Results

With all the hypothesis described above and recapitulated in Table 3.16, we calculate the Net Present Value (NPV) of the project for an installed capacity between 0.5 and 2.5 kWp. Results appear in Fig. 3.14.

It results that for a 0.5 kWp capacity the NPV is negative, it means that the installation is not profitable. We could expect this result as for such a low capacity the fixed costs of installation cannot be balanced by the savings on the electric bills because the energy production is too low. Then, the important information which provides this graph is that the capacity that optimizes the profitability is 1.5 kWp; even if the difference between 1.5 and 2 kWp is very slight. To confirm this result we evaluate two other project indicators: the Internal Rate Return (IRR)—Fig. 3.15 —and the payback (time to get the investment back) (Fig. 3.16).

Obviously, the 1.5 kWp capacity maximizes the IRR and minimizes the payback, but what it is interesting to note is that, again, the difference with 1 or 2 kWp capacity is not so important. Indeed, the installation will be amortized after 14 years in the best case, and after 15 years in the two other cases, this is not a very significant difference.

As a conclusion, we can say that the most profitable capacity to install for a typical household in Madrid area is 1.5 kWp, even if capacity of 1 and 2 kWp are

Fig. 3.15 IRR in function of
the capacity

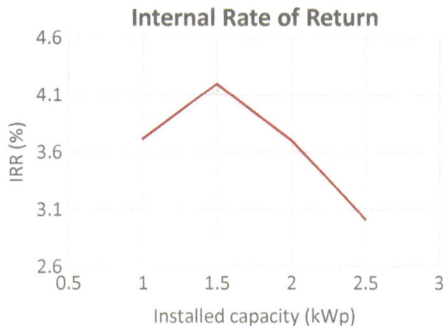

Fig. 3.16 Payback in
function of the capacity

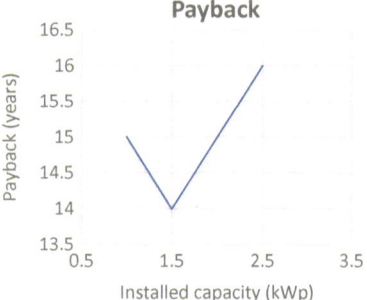

also totally acceptable. Installing a capacity inferior to 1 kWp is not profitable
because the energy production is too low to balance the fixed costs of the installation. Installing more than 2 kWp neither is profitable because there are too many
energy losses as the modules produce way more than what the household really
needs.

3.3   Sensitivity Analysis

In the anterior section, we displayed the results for a basic scenario, that is to say
without variation of the electricity price, with low prices for the equipment and with
the irradiation available in Madrid area. In this section we are going to carry out
several sensitivity analysis on these parameters in order to determine to which one
the model is the most sensitive.

3.3.1 Cost of Equipment

In Sect. 3.1.5 we defined the cost of equipment considering two scenarios, one with low prices and another one with high prices. Until now, all the results displayed were obtained with the hypothesis of low prices. We will here see what occurs with the high prices.

Before starting it is worth reminding that the HP scenario deals with the following price function: 2500 + 250 * N where N is the number of modules needed.

Table 3.17 shows the NPV values for the different installed capacity. Without surprise, if we draw the curve NPV against the capacity we will obtain the same profile than in Fig. 3.14: the 1.5 kWp capacity is the one that optimizes the profitability of the installation. However, all the NPV are negative, its means that the PV system will not be profitable in any case.

We said that this HP scenario stood for one client that has no specific knowledge in solar energy and would buy an all-included package so as not to spend time and energy to buy, install and maintain it by himself. And the conclusion is that such an investment is not profitable under the hypothesis we made, that is to say without energy storage and without selling the surplus to the grid.

Figure 3.17 displays the NPV in function of the cost of equipment, and therefore the limit price for which the system is profitable. Of course, the curves are affine and the slope does not depend on the installed capacity considered. For a 1.5 kWp installation, the limit price is 3187 €, it corresponds to an installation price of 2.12 €/Wp. This price is superior to the range price given in Table 3.10, which said that the final price of a PV installation is currently between 0.9 and 2 €/Wp. Consequently it confirms that with a price in this range, the system would be then profitable.

3.3.2 Electricity Price

3.3.2.1 Different Evolution Scenarios

As we saw in Sect. 3.1.1.3, prices of electricity are very volatile and impossible to predict on the long term. Even if it exists economic models that try to make forecasts for the next years, none of them is able to tell what will be the retail electricity price in 20 years. This is why we chose in the previous profitability

Table 3.17 NPV in function of the installed capacity considering the high prices scenario

	0.5 kWp	1 kWp	1.5 kWp	2 kWp	2.5 kWp
NPV (€)	−2195	−833	−813	−1135	−1575

Source Own elaboration

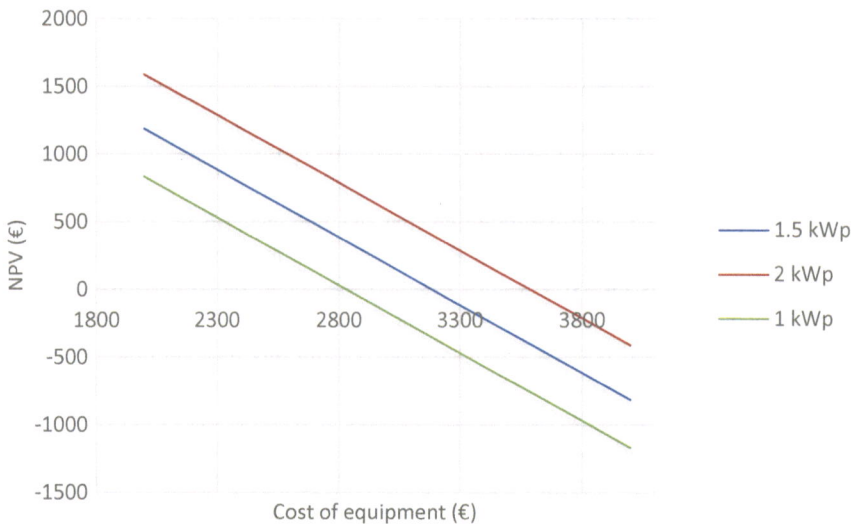

Fig. 3.17 Sensitivity of the NPV to the cost of equipment. *Source* Own elaboration

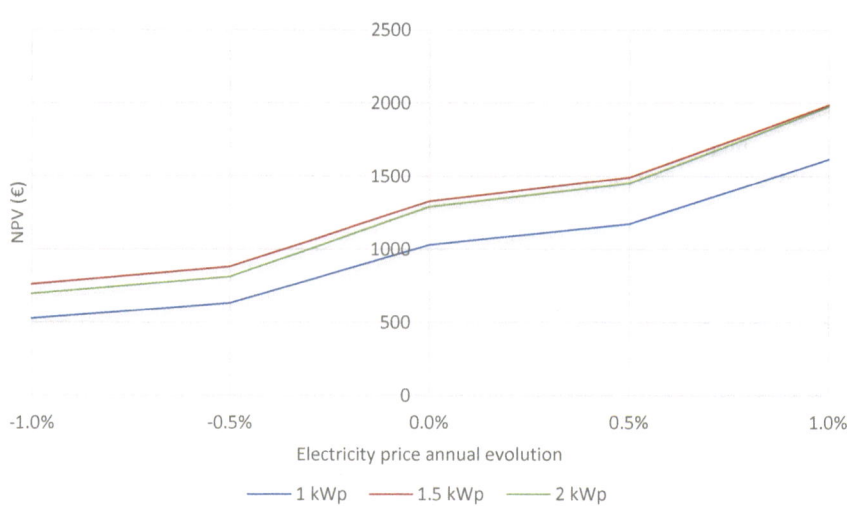

Fig. 3.18 NPV sensitivity to the electricity price's evolution. *Source* Own elaboration

evaluation to consider it invariable over the 25 years of the study. However, in this
section we will see what could be the consequences of a variation in the electricity
price over time. For this, we will consider two basic scenarios: a first one in which
the price raises 1% each year and a second one in which the price falls 1% each
year. None of these scenarios are impossible. On one hand, during the last decades
electric prices in Europe tended to increase. According to Eurostat [1], the average

end user price in EU-28 passed from 0.18 €/kWh in 2008 to 0.206 €/kWh in 2016, which represents an annual augmentation of 3.3%. But on the other hand, the recent entrance of many renewable energies in the energy market disrupts it totally, and forecasting a drop in the prices is not something improbable.

Figure 3.18 shows the NPV (calculated with the same hypothesis as in the anterior section) in function of the electricity price annual evolution and for the 3 meaningful installed capacity.

For each capacity the shape of the function is the same, it means that the sensibility to the electricity price does not depend on the capacity of the installation. Consequently, there is no capacity that could best balance the electricity price volatility than another. Moreover, this graph tells us how sensible the model is to energy price variation. Indeed, if it annually raises 1%, the NPV increases 35% in comparison with a fixed price. This number is significant, as for an investor it is not the same making an investment and getting back 35% more or less money. Consequently, it illustrates one of the major difficulty for solar energy development: it deeply depends on prices of the energy market, and nobody can really tell how they are going to progress. So it keeps being a relatively risky investment and some people might be afraid of that.

3.3.2.2 Different Electrical Tariffs

As explained in Sect. 3.1.1.4, it exists in Spain different kind of electrical tariffs: the classical we have used until now (fixed retail price 0.162 €/kWh), and the 2.0 DHA and 2.0 DHS one with which the client pay its electricity at a different price in function of the hour of the day. In this paragraph we are going to study which is the tariff that best optimizes the profitability of a solar installation. For this we have simply calculated the project profitability indicators for the 2.0 DHA and DHS tariffs and compared them with those previously obtained for the classical contract.

To calculate the indicators we used the same model than before, we just have to separate the energy produced and used during peak hours from the one produced and used during off-peak hours, and apply a different price to them. For the 2.0 DHA tariff, there are only two different time periods: peak hours from 1 pm to 11 pm and off peak hours the rest of the time. This is for summer, in winter the changes in prices occur at 12 am and 10 pm respectively. For the 2.0 DHS tariff there are three periods identical during the year:

– Peak hours from 1 pm to 11 pm
– Off-peak hours from 11 pm to 1 am and from 7 am to 11 am
– Super off-peak hours from 1 am to 7 am

However, just two of these periods will take part in our model. Indeed, there is no solar energy that can be produced by night, so the super off-peak price will not be used. Consequently, regarding our study, the unique difference between

Table 3.18 Retail prices in €/kWh in function of the time periods for 2.0 DHA and DHS tariffs (marginal cost)

	Peak hours	Off-peak hours	Super off-peak hours
2.0 DHA	0.19	0.09	–
2.0 DHS	0.2	0.11	0.08

Source Own elaboration

Table 3.19 Distribution of the electric consumption between peak and off-peak hours for clients in Madrid area subscribed to the DHS tariff (in kWh/year)

	0.5 kWp	1 kWp	1.5 kWp	2 kWp	2.5 kWp
Consumption peak hours	355.7	606.5	695.5	735.6	761.9
Consumption off-peak hours	416.8	702.5	774.8	796.9	812.3

Source Own elaboration

subscribing a 2.0 DHA or a 2.0 DHS contract is the electricity price during peak or off-peak hours. Table 3.18 displays the corresponding prices.

A PV system produces electricity only during sunned hours, that is to say approximately between 7 am and 8 pm in summer and between 8 am and 6 pm in winter. Therefore, roughly half of the production is made during peak hours and the other half during off-peak hours (Tables 3.19 and 3.20 figure the exact number for each installed capacity considered). For the electricity produced during peak hours the householder will save more money than with a traditional contract, but during off-peak hours they will save less. The question is then to know if the higher retail price during peak hours compensates the lower one during off-peak hours.

It turns out that installing a PV system in a house where the electricity contract is with time discrimination is less profitable. Indeed, Table 3.21 shows the three financial indicators for each type of contract and we see that for the 2.0 DHA contract the NPV is approximately two times lower and for the 2.0 DHS one it is nearly 35% lower than for the normal tariff.

We could expect these results as 2.0 DHA and DHS tariffs are made for clients who consume more energy during the night or early in the morning. On the contrary, to make profitable a PV system without storage or without resales to the grid, the consumption must be the highest possible during sunned hours. For example, tariffs with time discrimination are highly recommended for households with an

Table 3.20 Distribution of the electric consumption between peak and off-peak hours for clients in Madrid area subscribed to the DHA tariff (in kWh/year)

	0.5 kWp	1 kWp	1.5 kWp	2 kWp	2.5 kWp
Consumption peak hours	403.3	687.1	782.0	822.1	848.4
Consumption off peak hours	369.2	621.9	688.3	710.4	725.8

Source Own elaboration

Table 3.21 Financial indicators in function of the tariff subscribed

	NPV (€)			IRR (%)			Payback (years)		
	Normal tariff	2.0 DHA	2.0 DHS	Normal tariff	2.0 DHA	2.0 DHS	Normal tariff	2.0 DHA	2.0 DHS
0.5 kWp	−614	−956	−819						
1 kWp	1033	465	694	3.72	1.79	2.59	15	19	17
1.5 kWp	1330	717	977	4.2	2.39	3.18	14	17	15
2 kWp	1290	648	931	3.71	1.97	2.76	15	19	16
2.5 kWp	1131	509	801	3.02	1.43	2.1	16	20	18

Source Own elaboration

electric car as in this case the car would be charged during the night when the electricity is the cheapest.

3.3.3   Sensitivity to Solar Irradiation

The previous results were obtained considering the irradiation in Madrid. In this paragraph we will study the influence of irradiation on the installation's profitability by implementing the model in Barcelona and Seville. Of course, as Seville is sunnier than Madrid we will find a higher profitability there, but the idea is to see if this higher profitability could have a significant impact on a customer's decision to invest or not.

First, Table 3.22 shows the number of kWh per year that a householder can save on his electrical bills depending on his geographical position. It turns out that there is no difference between Barcelona and Madrid, and that there is a slight difference between Seville and Madrid. In Seville, in function of the installed capacity, a client can use between 1.6 and 4% more solar energy each year. It represents only between 4 and 9 € saved on the bill, which is not really significant.

Then we calculated the financial indicators, which appear in Table 3.23, for Seville area. As we could expected, the results are quite similar to those we get for Madrid area. In the best case, which corresponds to a 1.5 kWp installation, the NPV is worth 1463 € while it was 1330 € for Madrid, the IRR goes from 4.2 to 4.57% and the payback is 14 years in both cases. Again, the difference is not significant and we can conclude that the geographical position within Spain would not

Table 3.22 Useful energy in function of the geographic position

Useful energy (kWh)	0.5 kWp	1 kWp	1.5 kWp	2 kWp	2.5 kWp
Seville	812	1362	1507	1562	1596
Madrid	772	1309	1470	1537	1571
Barcelona	752	1300	1465	1531	1573

Source Own elaboration

Table 3.23 Financial indicators for a PV system installed in Seville area

	NPV	IRR (%)	Payback
0.5 kWp	−473.5		
1 kWp	1123.0	4.33	14
1.5 kWp	1463.5	4.57	14
2 kWp	1182.0	3.43	15
2.5 kWp	901.0	2.45	17

Source Own elaboration

influence someone's choice to buy and install a solar panel in his house. This is interesting because it is not an intuitive statement. As Northern Spain is seen much more rainy and cloudy than the Southern part, people could think that it is not worth installing a PV system in the North. The previous results prove that they are wrong, the system profitability is less than 8% lower in the North than in the South, and considering the small amount of money which is at stake it is not a noteworthy difference.

3.4 Levelized Cost of Electricity

3.4.1 LCOE in 2016

As we saw it in Sect. 2.2 of the literature review, a powerful tool to evaluate grid parity is the Levelized Cost of Electricity (LCOE). It is worth reminding that the LCOE basically corresponds to the production cost of 1 kWh of electricity, in our case, using a PV system. To calculate the LCOE we used the model presented by Hernandez-Moro and Martinez-Duart [15], who themselves relied on the work of Branker, Pathak and Pearce [16].

The concept of this calculation is very simple, it is dividing the total costs generated by the equipment over its lifespan (25 years here) by the energy produced during this same period. As the expenses and sales revenues that occur in a future time have to be accounted for the present time value of money, we have to use a discount rate r that means to evaluate the present value of the cash flow. Then, we obtain the following expression:

$$\text{LCOE} = \left(\sum_{n=0}^{N} \frac{\text{Costs}_n}{(1+r)^n} \right) / \left(\sum_{n=0}^{N} \frac{E_n}{(1+r)^n} \right) \tag{3.2}$$

where E_n is the energy produced by the PV system during year n, in our case it corresponds to what we called useful energy; N stands for the number of year that are considered for the study (N = 25 here); and Costs_n refers to every cost related to

Table 3.24 Discount rate used to calculate the LCOE in the literature

Paper	Discount rate (%)
International Energy Agency [17]	10–12
Breyer and Gerlach [18]	6.40
Bhandari and Stadler [14]	4
Mayer, Philipps et al. [19]	5–10

Source Own elaboration

the equipment in each year, it includes the initial investment and the maintenance costs.

We already have determined the different costs and the energy produced. The unique parameter that is still unknown is the discount rate. In the literature discount rates between 4 and 12% are assumed (see Table 3.24) as solar investments are considered relatively risky. In our case we can reasonably choose a 5% discount rate because the investment is for small installations and consequently does not represent a big amount of money even for a particular.

The choice of the discount rate is quite important because it will greatly influence the value of the LCOE as you can see in Fig. 3.19. In this graph, the retail electricity price is also represented in order to determine for which discount rate grid parity is reached. It turns out that for a 5% discount rate, the LCOE is still higher than the end user price, it means that there is no grid parity. Grid parity is reached only for discount rate below 4%, this is coherent with our previous results as we obtained an IRR of 4% in the best cases.

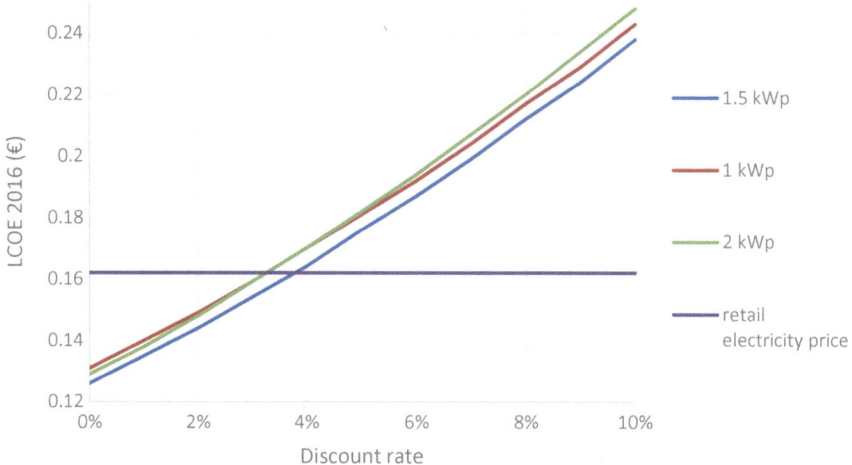

Fig. 3.19 LCOE in function of the discount rate (with data from Madrid). *Source* Own elaboration

3.4.2   LCOE Forecasts

We saw that under our hypothesis grid parity is still not reached in Spain. Indeed, the solar LCOE is currently at 0.176 €/kWh whereas the marginal cost of grid electricity is at 0.162 €/kWh. But, as we saw in the literature review, costs equipment for PV technology are decreasing since its apparition on the market and will keep doing so. Then, the idea is, considering this decline, to forecast when the grid parity will be effective.

For this, we are going to use the projection model from Biondi and Moretto (2014) [20] which we have already presented in Sect. 2.2 of literature review. In this model (see Eq. 2.8), the LCOE at time t depends on the initial value of LCOE and on an exponential factor (see Eq. 2.7) relative to the learning and the growth rate of solar system. Accordingly to what we said in the literature review, we will consider a learning rate of 20%. The growth rate is trickier to evaluate, this is why we will consider two scenarios [20]: an optimistic one with 18% GR and a conservative one with 20% GR. The results regarding these two scenarios are plotted in Fig. 3.20. We notice that the LCOE is decreasing in both cases quite quickly. With the conservative scenario, grid parity is reached within 3 years, while for the optimistic one within 2 years. It means that, in any case, grid parity is almost there in Spain. Then, if we look at the long-term results, we see that the LCOE is supposed to take very low values: in 2030, 0.078 €/kWh with the optimistic scenario and 0.112 €/kWh with the conservative one. It is difficult to say if these values are realistic or not, but surely this model is more accurate for short-term previsions because it depends mostly on the learning and the growth rates of the solar market which is likely to vary during next years.

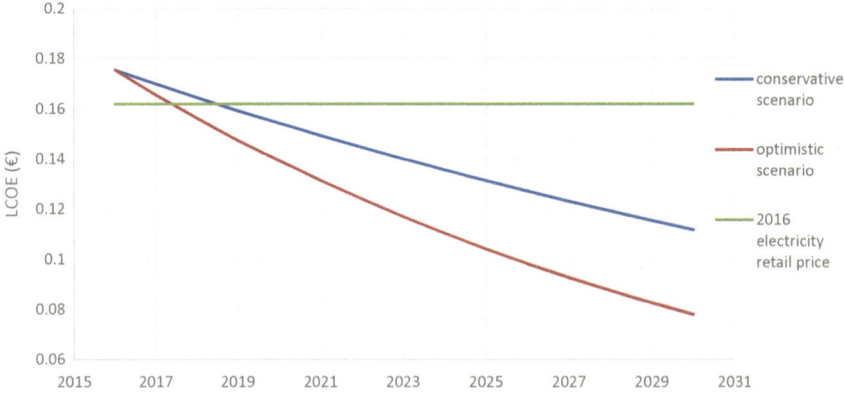

Fig. 3.20 LCOE forecasts considering a conservative or an optimistic scenario. *Source* Own elaboration

3.5 Main Findings

According to our model, there is still no grid parity for solar energy in Spain, and it will happen within 2 or 3 years. Nevertheless, it is worth reminding that it is a model quite conservative, therefore it is not surprising that some papers consider grid parity already reached (for instance Breyer and Gerlach [18] quoted in the literature review). For the moment, the LCOE is worth 0.176 €/kWh, which is not far from the marginal cost of electricity from the grid 0.162 €/kWh. In the best case, installing a PV system has an internal rate of 4.57%, for an initial investment of 2450 €, it means it is equivalent to invest money with such a rate of interest. This optimum situation consists in being in the South of the country, choosing a 1.5 kWp capacity, benefiting from low prices on the equipment and hoping that electricity prices will not vary in the next years.

Besides, among these parameters to reach the best profitability, some are more important than others. Even if, being in a sunnier region helps to improve the benefits, it is not essential as we saw than the difference between Northern Spain and Southern Spain is really thin. However, if the national price of electricity were to change in the next years, even very slightly, these results would be totally changed, in better if it were to raise or in worse if it were to decrease.

Finally, we can conclude that the domestic use of solar panels in Spain, without energy storage and only for self-consumption, is profitable but in a limited way. The 4% profitability, for an initial investment quite low, would not always be enough to balance the time and the implication necessary for someone wishing to install a PV system for his dwelling. We can imagine that people who actually take the plunge in solar energy are also motivated by the ecological argument (which will be detailed in Chap. 5).

References

1. Eurostat Statistic Explained. http://ec.europa.eu/eurostat/statistics-explained/images/2/29/Electricity_and_gas_prices%2C_second_half_of_year%2C_2013%E2%80%9315_%28EUR_per_kWh%29_YB16.png
2. Sancha Gonzalo JL (2014) El Sistema Eléctrico Español (IV). *Anales de mecánica y electricidad/septiembre-octubre 2014*, pp 23–33
3. Energía info, web page aiming at giving comparative information on Spain's electrical and gas tariffs. http://www.energia-info.es/comparativa-tarifas-luz/
4. Joint Research Centre, European commission. http://re.jrc.ec.europa.eu/pvgis/apps4/pvest.php
5. Mulvaney D (2014) IEEE spectrum, 13th Nov 2014
6. Eco2mix, data base of RTE. http://www.rte-france.com/fr/eco2mix/donnees-en-energie
7. http://cenergypower.com/blog/string-vs-central-inverters-choosing-right-inverter/
8. Guisado-Falante JM, Lillo-Bravo I (2016) Bachelor's thesis "Viabilidad del almacenamiento energético en instalaciones fotovoltaicas de autoconsumo". University of Seville
9. Pablo Eguia, Esther Torres and Javier García, July, 28th 2016. Proyecto Gedisper. Informe final
10. https://www.wec-indicators.enerdata.eu/household-electricity-use.html

11. Disposición 3069 del BOE número 68 de 2016, document edited by the Spanish minister of presidency and territorial administrations
12. Online data base of REE. https://demanda.ree.es/movil/peninsula/demanda/tablas/2016-01-14/3
13. Daniel Lugo Laguna's master thesis: "Gestión de los picos de potencia eléctrica a nivel doméstico mediante almacenamiento energético en baterías"
14. Bhandari R, Stadler I (2009) Grid parity analysis of solar photovoltaic systems in Germany using experience curves. Sol Energy 83:1634–1644
15. Hernández-Moro J, Martínez-Duart JM (2013) Analytical model for solar PV and CSP electricity costs: present LCOE values and their future evolution. Renew Sustain Energy Rev 20:119–132
16. Branker K, Pathak MJM, Pearce JM (2011) A review of solar photovoltaic levelized cost of electricity. Renew Sustain Energy Rev 15:4470–4482
17. International Energy Agency. World Energy Outlook (2016)
18. Breyer C, Gerlach A (2013) Global overview on grid-parity. Prog Photovoltaics Res Appl 21:121–136.
19. Kost C, Mayer JN, Thomsen J, Hartmann N, Senkpiel C, Philipps S, Schlegl T (2013) Levelized cost of electricity renewable energy technologies. Fraunhofer Institute for Solar Energy Systems ISE
20. Biondi T, Moretto M (2015) Solar grid parity dynamics in Italy: a real option approach. Energy 80:293–302

Chapter 4
International Comparison

In this section we will carry out the study in other European countries so as to determine in which country grid parity is more likely to be reached.

The first problem we faced is that the daily curve of the domestic electrical consumption is not publicly available in most of the countries. Consequently, we decided to limit our study to Mediterranean countries and to make the hypothesis that the shape of the load curve in Spain is acceptable for them. Indeed, what greatly characterizes its profile are the peak hours which depends on the climate (necessity for heating or for air-conditioning) and on the duration of the day (if the sunset comes early the needs in lighting are higher); and we can consider that these parameters are quite similar for all the Mediterranean area. However, these countries have different standards of living that influence the electric consumption. For instance, Greece is poorer than Spain, this is why, on average, Greek households have less domestic equipment (TV, dishwasher, video games console…) than Spanish one, and therefore they consume less electricity. Moreover, the load curve is also influenced by the heating system most used in the country, as in some gas heating systems are more popular than electrical one for example. To take that into account, each curve will be weighted with the average electric consumption of its respective country.

Furthermore, in some countries that have a large difference of latitude between its more Northern point and Southern one, we will deal with several areas of the territory (as we did for Spain). So the geographic areas which will be studied are the following ones: Lisbon, Marseille (South of France), Milan, Naples, Malta and Athens. To carry out the study the same methods as for Spain are used, this is why we will not detail again the process but just reveal the most relevant results.

© The Author(s), under exclusive licence to Springer Nature Switzerland AG 2019   51
Á. Arcos-Vargas and L. Riviere, *Grid Parity and Carbon Footprint*,
SpringerBriefs in Energy, https://doi.org/10.1007/978-3-030-06064-0_4

4.1 Lisbon, Portugal

In Sect. 3.1.1.2, we saw that the difference between the marginal cost and the average cost for the end user electricity price was slightly less important in Portugal than in Spain. The marginal cost then calculated was 0.17 €/kWh, but just as we did for Spain, we will use here more actualized tariffs. Table 4.1 displays the prices offered by the main companies in the country, the average price obtained with them will be therefore taken for the study. Pay attention that in the table the VTA is already included in the prices.

Then, the two other components of the model that need to be modified are the average annual domestic consumption in electricity and the daily irradiation. The second one is obtained again thanks to the tool from the JRC European commission [1] and is very similar to the irradiation in Spain. In Seville, as well as in Lisbon, a 1 kWp capacity installation can produced up to 1623 kWh per year (this is not useful energy but just produced energy). However, the monthly production differs a little bit between the two areas, in Seville it is more homogeneous over the year, whereas in Lisbon there is more difference between summer and winter. We will see later that it plays an important role in the results. Besides, the average electrical consumption is slightly lower in Portugal than in Spain, probably because of the lower standard of living, it is approximately 3000 kWh per year per household [2].

The results are comparable to those obtained for Spain as the capacity that optimizes the NPV is also 1.5 kWp. Nevertheless, the profitability of the system will be lower. We can see in Table 4.2 that in the best case the NPV is worth 983 € and the IRR 3.2%, which is respectively 26% less and one point less than with the same installation in Madrid.

This lower profitability can be explained by a higher amount of energy losses in Portugal especially in summer. Indeed, during the most sunned months the solar

Table 4.1 Marginal cost of electricity (taxes included) offered by different Portuguese companies

	Marginal cost (€/kWh)
EDP eletricidade	0.164
Elusa BTN	0.158
GALP	0.166
Iberdrola	0.173
Luzboa	0.156
YLCE	0.155
Goldenegy	0.164
Enat Trinca	0.162
Audax	0.158
Energia	0.159
Endesa	0.161
Average	0.161

Source lojaluz.com [8] and own elaboration

Table 4.2 Financial indicators for a PV installation in Lisbon

	NPV	IRR (%)	Payback
0.5 kWp	−491.0		
1 kWp	853.5	3.13	15
1.5 kWp	983.3	3.20	15
2 kWp	870.0	2.59	16
2.5 kWp	692.4	1.91	19

Source Own elaboration

panels production is higher in Lisbon than in Spain (even than in Seville), whereas the domestic demand is a bit lower. Therefore, there is much more energy that is produced but cannot be consumed, and these losses cannot be balanced by the winter months since the production is then lower than in Spain. For instance in July and August the losses in Portugal for a 1.5 kWp are up to 48% against 43 and 40% in Seville.

Under these hypothesis, a 5% discount rate and a 1.5 kWp capacity, the LCOE is worth 0.187 €/kWh which is obviously higher than the retail price. Grid parity is not yet there, and according to the forecast formula (Eq. 5), it will be reached in 2021.

4.2   Italy

As Italy stretches above more than 1000 km from North to South, the analysis will be made considering two different areas, Milan area in the North and Naples area in the South. Italy has a perfil quite similar to the Spanish one: a Mediterranean country where the electricity is quite expensive, the solar irradiance relatively abundant and the electrical consumption not so high. Consequently the model will provide us results comparable to the previous ones.

Italy has an average domestic electric demand a little lower than Spain, it is worth 2700 kWh/year [3, 4]. But as the solar irradiance is also slighty inferior, the energy losses resulting are identical, at least between Milan and Seville. Table 4.3 displays the results for a 1 kWp installation, we can see that if we consider only the local irradiation, Seville and Naples have a similar solar potential which is superior

Table 4.3 Solar production and losses in Italy for a 1 kWp installation

	Solar production (kWh/year)	Losses[a] (%)	Useful energy (kWh/year)
Milan	1300	18.40	1063
Naples	1510	24.40	1144
Seville	1620	16.10	1362

Source Own elaboration
[a]Losses = energy that is produced but not consumed when the demand is inferior to the production

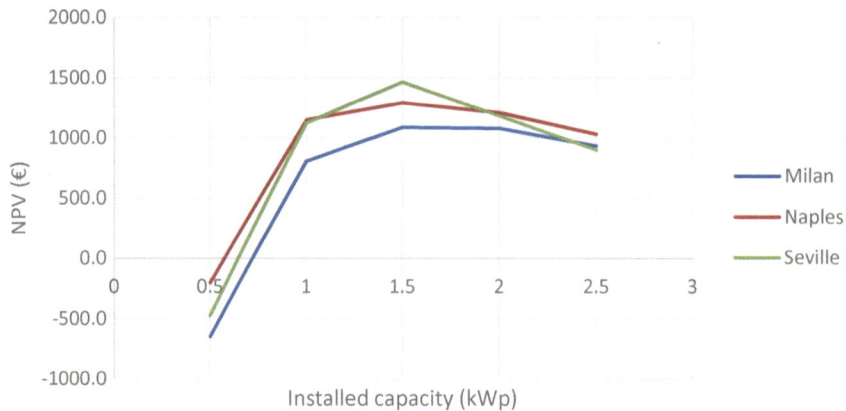

Fig. 4.1 NPV comparison in function of the area and the installed capacity. *Source* Own elaboration

to the one in Milan. However, the losses are higher in Naples than in Seville, consequently the useful energy obtained is more comparable between Milan and Naples than between Naples and Seville.

Concerning the electrical tariff, Italy is one of the most expensive country in Europe (see Fig. 3.2). In 2012 the marginal cost was worth 0.22 €/kWh [5], but since then it has slightly decreased, we will therefore consider a 0.19 €/kWh retail price. With all these hypothesis we can calculate the NPV for each installed capacity to determine which one optimizes the profitability. Results are shown in Fig. 4.1, it turns out that in Naples as well as in Milan, it is again the 1.5 kWp capacity that is the most profitable. Nevertheless the difference with 1 or 2 kWp capacity is really tiny, this is why we can say that installations from 1 to 2 kWp are acceptable in Italy. Moreover, we see without surprise that a PV system is more cost-effective in the South of Italy than in the North. But, like we said it for Spain, the difference between Northern and Southern areas is not so significant, and the conditions of investment are then quite similar in both cases.

Concerning the IRR and the payback, they are analogous to the ones obtained in Spain. In the best case, the IRR is worth 4.1% in Naples and 3.5% in Milan and the payback respectively 14 and 15 years.

Finally, the LCOE with a 5% discount rate is valued at 0.216 €/kWh in Milan and 0.207 €/kWh in Naples, it means that grid parity is still not there. If we want grid parity to be reached right now we have to consider a 3 and 3.5% discount rate respectively for Milan and Naples. The forecast model tells us that without modifying the discount rate it will be reached in 3 or 4 years.

Table 4.4 Details of the variable component of an electric bill in France, source: a 2016 EDF bill

	€/kWh
Energy cost	0.0932
TCFE tax	0.009
VAT (19.6%)	0.0200312
Marginal cost	**0.122**

Source Own elaboration

4.3 Marseille, Southern Part of France

As we explained in the introduction of Chap. 4, we do not have access to the daily perfil of domestic electric demand in France, and therefore we chose to take the Spanish one and weight it with the national annual consumption. This perfil is valid only for countries with a Mediterranean climate, this is why for France we will just consider its Southern part, taking data from Marseille area.

Due to its high standard of living, France is a large consumer of electricity, the average annual consumption for a dwelling is up to 4760 kWh. Consequently the demand curve will be higher than in the precedent cases and we will obtain losses much lower. Concerning the price of electricity France is a special case among Mediterranean countries. Thanks to the high rate of electricity coming from nuclear, the end user price is much lower than in the nations we studied until now. However, meanwhile electricity is going cheaper in Italy or in Spain for instance, in France it is going more expensive. In 2012, according to Sancha Gonzalo [5], the marginal cost was worth 0.118 €/kWh, now it is worth 0.122 €/kWh (see details in Table 4.4).

The first difference with the countries we studied before is that because of the higher needs in electricity, the range of installed capacity acceptable changes: it goes from 1–2 to 1.5–2.5 kWp. Table 4.5 features the exact values of each financial indicator. We see that again the three of them have the same order of magnitude than Spain, Portugal or Italy. But in this case, the capacity which optimizes the profit is the 2 kWp one.

It turns out that the low cost of electricity is balanced by a superior amount of useful energy. In Table 4.6 you will find the values of the losses in comparison with the case of Seville. We see that despite a lower solar irradiance but thanks to losses significantly lower as well, in France a PV system can produce more useful energy

Table 4.5 Financial indicators for a PV installation in South of France

	NPV	IRR	Payback
0.5 kWp	−1357.0		
1 kWp	390.0	1.51%	20
1.5 kWp	1168.0	3.74%	15
2 kWp	1334.0	3.82%	15
2.5 kWp	1261.0	3.33%	15

Source Own elaboration

Table 4.6 Solar production and losses in Marseille and Madrid for 1.5 and 2 kWp installations

	Solar production (kWh/year)	Losses (%)	Useful energy (kWh/year)
Seville—1.5 kWp	2440	38.10	1507
Marseille—1.5 kWp	2250	16.00	1892
Seville—2 kWp	3250	51.90	1560
Marseille—2 kWp	3000	31.60	2057

Source Own elaboration

than in Spain. Nevertheless as we saw it above the profit is not superior because of a cheap electricity.

Finally the LCOE under these hypothesis is evaluated at 0.136 €/kWh with the typical 5% discount rate. Consequently, the situation is the same as in Italy or Spain, grid parity is not reached yet but according to the forecast model it will be so in 3 or 4 years.

4.4   Malta

Malta is a case quite different from the other countries studied in this section as it is an island totally dependent on fuel and on importation to supply its inhabitants with electricity. Consequently the electricity is quite expensive there, but thanks to a VAT very low, the price does not go upper than in Spain or in Italy for instance. Another particularity of the country is that for particulars it exist two distinct tariffs, the residential one which is the cheapest and the domestic one which is significantly more expensive. The first one is the more widespread and applies to households which are the main home of its occupants. The second one was created for secondary residences, but actually many people renting a house and living there at long term have to pay this price as well, especially foreigners. Indeed, the residential tariff is applicable to maximum two homes with the same landlord. It means that a landlord owning several homes and renting all of them must make pay a part of his clients the domestic tariff. The marginal cost of the residential tariff is 0.130 €/kWh whereas for the domestic one it is 0.167 €/kWh, a difference of 25% that can play a key role when it comes to take the decision of investing or not in a solar system.

As the residential tariff is the more common on the island, we will use this price to determine whether grid parity is reached or not in Malta. Nevertheless we will also display the results under the hypothesis of the domestic tariff to see the difference between both cases.

Thanks to its geographic situation, the solar irradiance is quite important in Malta: with a 1 kWp installation up to 1680 kWh can be produced per year, which is more than in Seville (1620 kWh/year). Moreover the average domestic demand in electricity is high, it is worth 4000 kWh/year for an average home. This data was

Table 4.7 Daily electric demand in Malta in function of the kind of dwelling, source [6]

	Apartments	Maisonettes	Terraced Houses	Villas
Daily consumption in kWh	9.86	10.76	11.02	12.72

Source Own elaboration

calculated from a study written by Galea [6] in which he determines the daily consumption of different types of house present in Malta (Table 4.7).

We made the average of these four values, multiplied it by 365, and then obtained a yearly consumption of 4000 kWh. Malta is an island with an intensive touristic activity, many homes are rented all year to visitors who don't care about their consumption, and as well there are many expatriates with a very high standard of living. Therefore it is not surprising that the island has an average demand in electricity higher than Spain or Italy.

Table 4.8 displays the results of the model considering either the residential or the domestic tariff. With the residential tariff the values of the indicators are similar to those previously obtained for the other countries. The most profitable capacity is again the 1.5 kWp one, in this case the profit made is almost 4% and the investment is amortised after 14 years. However with the hypothesis of the domestic tariff the results are much better than in any other studied country. The profit goes up to 7.8%, the NPV is worth 2700 €, which is more than twice the value in the other case, and the payback is only 10 years.

Consequently, for a client subjected to the domestic tariff, installing a PV system is really profitable. The calculation of the LCOE in both cases confirms this result. It is worth 0.143 €/kWh with a 5% discount rate, it means that grid parity is already reached for the domestic tariff. Figure 4.2 shows the forecast of the LCOE for the next 8 years. We see that for the residential tariff grid parity will be reached within 3 years.

As a conclusion we can say that the situation in Malta about solar grid parity is similar to the one in the other Mediterranean countries if we consider the residential tariff. However, for the domestic tariff, PV energy is much more profitable, the conditions of investment are totally different. There is an important discontentment in Malta about the existence of these both tariffs, people arguing that they are living

Table 4.8 Financial indicators for a PV installation in Malta with residential or domestic tariff

	NPV		IRR		Payback	
	Residential	Domestic	Residential (%)	Domestic (%)	Residential	Domestic
0.5 kWp	−969.0	−280.0				
1 kWp	827.2	2108.5	3.05	6.96	15	11
1.5 kWp	1232.8	2709.6	3.92	7.78	14	10
2 kWp	1225.7	2780.6	3.54	7.27	15	11
2.5 kWp	1103.2	2703.3	2.95	6.53	16	12

Source Own elaboration

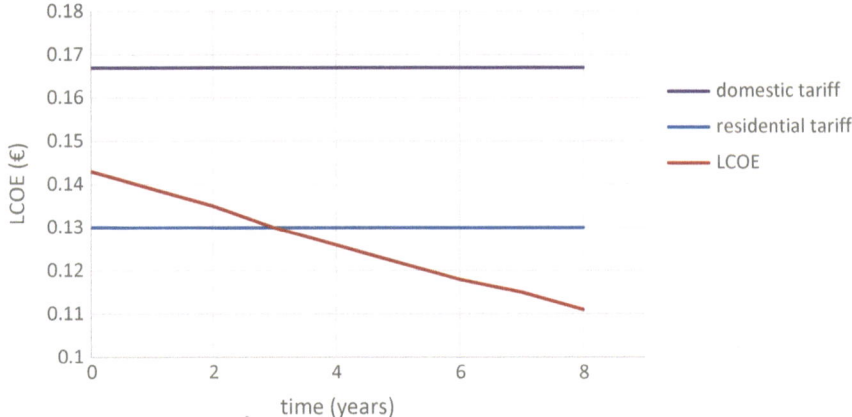

Fig. 4.2 LCOE forecast for Malta, discount rate 5%. *Source* Own elaboration

all year in the same house but have to pay the expensive price for their electricity, which is unfair. This study about PV energy illustrates the unfairness of these tariffs as being submitted to one or to the other can greatly change your energy policy.

4.5   Athens, Greece

The average electric demand in Greece is approximately 3000 kWh/year, which is the same as in Portugal. The end user electricity price is worth 0.176 €/kWh in 2016 [7], which is much more than a few years before. Because of the economic crisis, Greece knew a strong inflation and price of electricity did not escape from it as it raises approximately 50% between 2010 and 2016. We considered that the difference between marginal and average cost is similar than for the countries appearing in Table 3.2, and we applied the lower rate (−16%) to the average cost (0.176 €/kWh) to obtain an approximation of the marginal cost (0.147 €/kWh). With this method we get a marginal cost probably a little higher than it really is, which is in harmony with the conservative model we are dealing with.

The solar irradiation available in Greece is surprisingly a little lower than in Spain or Portugal. In the Iberian peninsula, with a 1 kWp installation you can produce up to 1620 kWh per year, whereas in Athens only 1530 kWh. Moreover the marginal cost of electricity is quite cheap (0.147 €/kWh), it means that the profitability of a PV system will be lower than in the other Mediterranean countries. Table 4.9 shows the financial indicators which are all very low. The most profitable capacity is the 1 kWp, but still the profit is very limited. Indeed, the initial investment (2200 €) is amortized only after 19 years, the NPV is worth 570 € and the IRR 2.15%. They are values probably too low to encourage people to invest into PV energy, even if theoretically they will not loose money.

Table 4.9 Financial indicators for a PV installation in Greece

	NPV	IRR (%)	Payback
0.5 kWp	−887.0		
1 kWp	567.4	2.15	19
1.5 kWp	286.1	1.00	21
2 kWp	177.9	0.57	23
2.5 kWp	−8.7		

Source Own elaboration

The LCOE is obviously higher than the retail price. It is worth 0.199 €/kWh for a 5% discount rate and according to the forecast equation it will take 10 years to reach the level of the retail price.

4.6 Summary on the International Comparison

For all the Mediterranean countries included in this study the solar grid parity is more or less at the same point. It will be reached within two to five years, except for Greece where it will take around ten years (see Fig. 4.3). This is for a 5% discount rate and it means that, if we consider a 3 or 4% discount rate, grid parity is already there. These last rates can seem low but actually they do not lack of sense for particulars. Indeed, common placements in bank account do not have much higher rates of profitability.

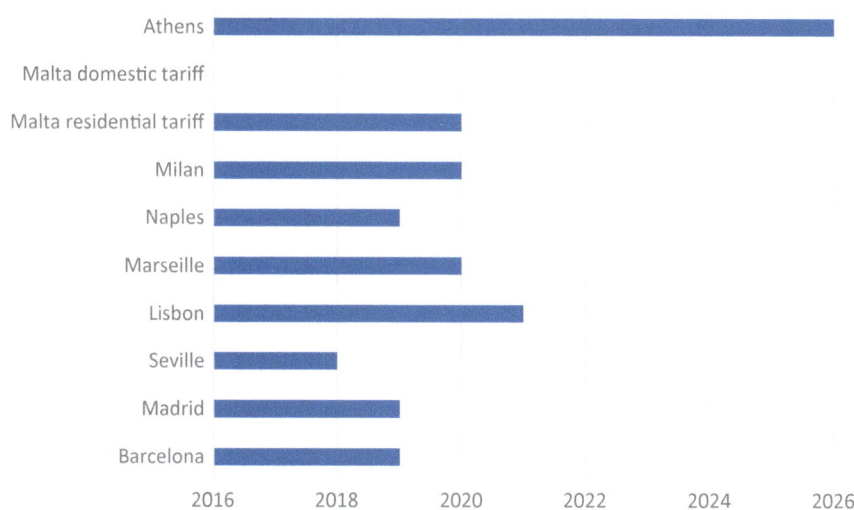

Fig. 4.3 Grid parity points for a 5% discount rate. *Source* Own elaboration

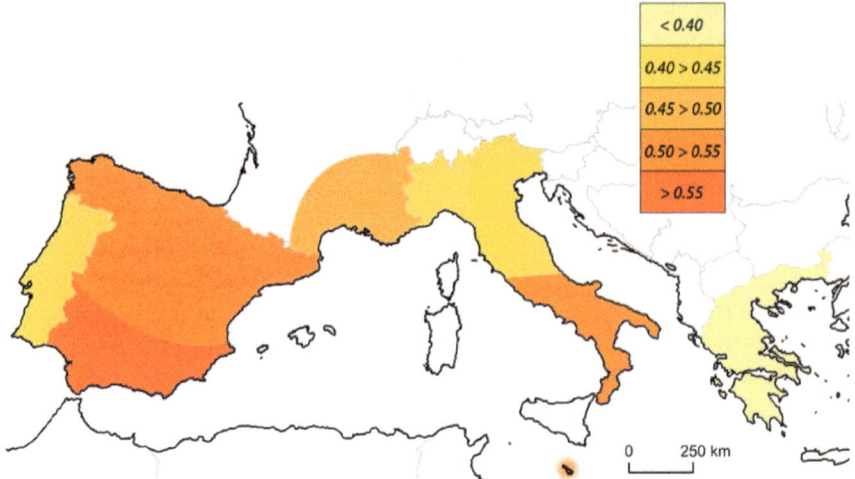

Fig. 4.4 Profitability indicator for solar energy: NPV/investment. *Source* Own elaboration

Another way of valuing the profitability of a project is to calculate the ratio NPV/initial investment. If this ratio is worth 0.5 it means the investor will get back, at the end of the project, 1.5 times the money he invested at the beginning. On Fig. 4.4, the values of this ratio are displayed for the Mediterranean countries. Without surprise, this ratio is the highest in Spain and the lowest in Greece. Nevertheless the difference is not so significant and investing in solar energy in Greece stays profitable. On the other hand, 0.55 is not a value so high for this kind of ratio, projects tend to be considered as really profitable when their NPV/Investment is closer to 1 or even superior to 1. But investing in a solar system is not only a question of economic profitability, there are also a lot of ecological and ethical parameters which are implicated in such a decision. We will tackle with these issues in the sixth chapter of this paper.

References

1. Breyer C, Gerlach A (2010) Global overview on grid-parity event dynamics. In: 25th European photovoltaic solar energy conference/WCPEC-5, Valencia, Sept 2010
2. Traça de Almeida A, Fonseca P. Residential monitoring to decrease energy use and carbon emissions in Europe. Intelligent Energy Europe. http://remodece.isr.uc.pt/workshops/portugal/REMODECE%20_Apresentacao_ISR-UC_29Set2008.pdf
3. Punti energía, web page aiming at giving comparative information on Italia's electrical and gas tariffs. http://puntienergia.com/guida/consumo-medio-energia-elettrica-famiglia
4. http://www.consulente-energia.com/af-quanti-kwh-consuma-una-famiglia-di-corrente-elettrica-consumo-annuo-di-elettricita-di-una-famiglia-media-italiana-di-2-3-4-persone.html
5. Sancha Gonzalo JL (2014) El Sistema Eléctrico Español (IV). Anales de mecánica y electricidad/septiembre-octubre 2014, pp 23–33

6. Galea J. Electricity consumption in households, European regional development fund, Malta 2007–2013. http://mra.org.mt/wp-content/uploads/2013/05/3850/NSO.pdf
7. https://www.statista.com/statistics/418083/electricity-prices-for-households-in-greece/
8. Loja luz, web page aiming at giving comparative information on Portugal's electrical and gas tariffs. http://lojaluz.com/preco-eletricidade-portugal

Chapter 5
Financial Analysis

Until now, the profitability of a PV system was evaluated according to the common project indicators, the Net Present Value and the Internal Rate of Return. They give really useful information and they have the advantage of being simple to calculate and to interprete. In short, they are a very good first approximation of the project's profitability and most of the time they offer enough information for the client to take the decision to invest or not. However, the NPV does not take into account the variability of some parameters and in some cases does not reflect all the possibility of investment. In particular, basing the decision of starting a project only on the value of its NPV is most of the time equivalent to forget the possibility of postponing the project and not to consider that the different variables may change in the course of the project's life time. For long-term investment with uncertain variables it can be a problem. The investment in solar energy corresponds to this case, as it is a project over 25 years with a great uncertainty on the evolution of electric prices. This is why the purpose of this chapter is to include in our financial evaluation hypothesis about changes in the electricity tariffs and see if it could change the decision about the investment.

This chapter is divided in two parts. First we will determine in which cases it could be advantageous to postpone the investment in the solar installation, and then which are the conditions that could justify the abandon of the project already underway. For both cases, we will consider the installation of a 1.5 kWp system in Seville area, which is one of the most profitable situations according to the previous analysis.

5.1 Is It Interesting to Postpone the Investment?

In this paragraph we will carry out a simple calculation that allows to know if it is better to postpone the beginning of the project or to invest right now. We make the hypothesis that the financial resources of the investor are illimited. In that case, it is

Á. Arcos-Vargas and L. Riviere, *Grid Parity and Carbon Footprint*,
SpringerBriefs in Energy, https://doi.org/10.1007/978-3-030-06064-0_5

worth postponing an investment if the cost of opportunity of the initial investment is superior to the first cash flow. This can be translated by the following equation:

$$n * r * I > \sum_{i=1}^{n} CF_i \tag{5.1}$$

where n stands for the number of years of postponement, r for the discount rate, I for the initial investment and CF_i for the value of the cash-flow in year i.

For our PV system, the initial investment is 2450 € and we consider a discount rate of 5% (see Sect. 3.4.1). The first year of the project, the cash flow is worth 232 €. Consequently, as $0.05 * 2450 = 122.5$, it is not advantageous to postpone the investment. In the same way, if we make the calculation for $n > 1$, we find that it is never worth waiting to start the project. This is because both the initial investment and the discount rate are quite low, therefore the cost of opportunity is weak as well and it never goes up the cash flows.

However, what does not consider this model is the advantage of postponing if we can have more information on the project and so reduce the risks. In our case, it would be really interesting if we could know for sure how is going to evolve the price of electricity. Unfortunately, even if we wait some years, there is very few probability of finding an exact forecast model for electrical tariffs. The only thing that could be done is to integrate in the calculation of the NPV a probability term which reflects the risk and the uncertainty of the investment. But this method requires much probabilistic data that we do not have, consequently it will not be performed in the present study.

5.2 Abandon of the Project

If during the course of the project the end user price of electricity suddenly became lower, the final profitability would be then way inferior to what was expected. Therefore, in this case, it could be interesting to stop the project, that is to say sale the panels and get back to buy all the electricity needed from the grid. The objective of this section is to determine under which conditions it is worth doing so. For that, we will use two criteria: in which year of the project electricity tariff starts to decrease and how important is the decline.

For the residual price of the solar modules, we make the simplifying hypothesis that they lose value in a linear way since their first year of running and until the end of its life (25 years). The initial price of the installation being 2450 €, it means each year it loses 98 € of its value. The other important hypothesis made is that once electricity price has started to fall, it does so in a regular way, suffering the same percentage decrease each year until the end of the project. To stay close to realistic and not catastrophic scenarios we will consider decreases only between −1 and −4% annually. In short, the question we are going to answer is "if in year x of the project electricity price starts to decrease y% each year, do we have to stop the project or to keep it going?"

The equation which models this problem is the following one:

$$\sum_{i=0}^{x} CF_i + P_x > NPV_{bis} \tag{5.2}$$

The left term represents the benefits if the project is interrupted, while the right-term stands for the case in which the project is maintained. NPV_{bis} corresponds to the new NPV taking into account the change in the electricity price, it is obviously lower than the initial NPV (1463 €). P_x is the residual price of the modules at year x of the project, and the sum of the CF_i represents the clash-flow accumulated before stoping the project. Consequently if this equation is true, it is more advantageous to stop the project. Note that the equation can be verified both terms being negative or positive. If they are negative it means that in both cases the investor will lose money, but the decision to uninstall and sale back the system allow to limit the losses. In the opposite if they are positive it means the project is profitable in any case, and that we have the opportunity to increase the benefits.

We made 96 different simulations, the year in which prices start to decrease varying from 1 to 24, and the annual rate of decline going from −1 to −4%. Results are displayed in Fig. 5.1. The vertical axis represents the difference $\sum_{i=0}^{x} CF_i + P_x - NPV_{bis}$, if it is positive it means it is better to stop the project, on the contrary, if it negative it is better to keep it going. We remind that the horizontal axis indicates the year in which the decline of prices begins.

The first striking point on this figure is the singular profile of the curves which all have two peaks at year 9 and year 19. But this phenomenon has a simple explanation, it corresponds to the years before the replacement of the inverter (its life

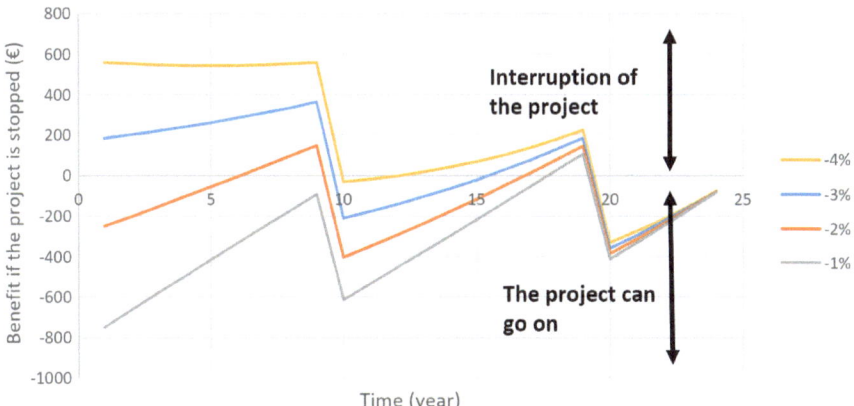

Fig. 5.1 Decision to abandon or not the project in function of the change in the electricity price. *Source* Own elaboration

time is only 10 years). As this component is the unique one which needs to be substituted and is quite expensive in comparison with the initial investment, it is logical that it has a great influence on the decision to stop or not the project. It turns out that just before buying a new inverter the benefit from interrupting the investment is much higher than a year after, because once a new inverter is bought, several years are necessary to amortize it.

More generally, we see that during the last five years of the panel's life time, whatever is the decline of the electricity it is not advantageous to sell back the solar modules because their residual value is then very low and the savings on the electrical bill stay always higher than this value. In the rest of the project the decision depends on the percentage of decrease. For −4%, it is almost always more interesting to stop the project, whenever is the change in the retail price. On the contrary, for −1%, it is not worth interrupting it, except in year 19 where the result is slightly positive. Furthermore, what we cannot see on Fig. 5.1 is if the investor will lose money even if it takes the decision indicated by this method. It turns out that there is a negative benefit in only 5 of the 96 cases. Of course these cases correspond to the major decrease in electricity price, precisely it occurs if it starts to fall at a 4% rate between year 1 and year 4, and at a 3% rate only the first year.

To sum up, if the electricity price starts to fall at a 1 or 2% rate, solar panel's user do not have to worry, they will not lose money in any case, and is almost never worth stopping and selling the installation. On the other hand, if this rate goes up to 3 or 4% (it could happen in situation of crisis, see what occurred in Greece after 2008) the situation could become trickier. In most cases, investors would have to stop their activity and sell the system in order to maximize their benefits.

Finally, we can say that investing in a PV system is not very risky as the probability to lose money is very low, however the time of amortization is quite long and the benefits will never be considerable.

Chapter 6
Carbon Footprint of Photovoltaic Energy

Until now we focused our study on the economic profitability of a PV system and it turned out that the investment was worth it for the finances of its owner. The next question is to know if this investment is also profitable for the Earth. As PV is most of the time considered as a green energy, we are tempted to answer a great yes to that question without thinking more. However, the reality is not so easy. It is true that once installed PV modules produce electricity without carbon emission, but their fabrication is a process quite complex that requires a significant quantity of energy and emits several GHG (Green House Gases). The objective of this chapter is therefore to quantify these emissions and to compare them to the emissions corresponding to the production of the national grid electricity. By doing so, we will be able to determine the energy payback time and the carbon footprint of our PV model.

6.1 How Much Carbon Emission Does a Solar System Save?

In this first section we are going to evaluate the quantity of carbon emission that is avoided by installing and using a PV system. In our model, this quantity corresponds to the carbon emission relative to the generation of electricity that is substituted by the production of the solar modules. The French and Spanish cases are going to be figured out as RTE[1] and REE[2] give public data about the instantaneous emission of CO_2 due to electricity production.

Consuming electricity from solar energy instead of from the grid reduces the marginal emissions corresponding to the last technology to enter the electrical generation system. Therefore, the non-emitted quantity of CO_2 results from the

---

[1]Réseau de Transport d'Electricité.
[2]Red Eléctrica de España.

© The Author(s), under exclusive licence to Springer Nature Switzerland AG 2019
Á. Arcos-Vargas and L. Riviere, *Grid Parity and Carbon Footprint*,
SpringerBriefs in Energy, https://doi.org/10.1007/978-3-030-06064-0_6

emissions rate of this last technology multiplied by the amount of energy produced by the PV system. The different technologies are entering in function of their running cost, the most expensive being the last one to be started. The ranking is the following, starting from the most expensive: fuel, gas and coal. Let us take an example to make it clearer.

Figures 6.1 and 6.2 present, for France and Spain on December 14th 2016, the distribution of the different technologies contributing to the generation of electricity. The technologies situated on the top of the curve are the carbon-emitting ones and are ordered in function of their entrance ranking (the most expensive technology is the last to enter the production system). Figure 6.1 reveals that around 70% of the French electricity is produced from nuclear energy and that the marginal emissions correspond to the utilization of fuel. Figure 6.2 shows that in Spain the distribution between all the technologies is much more homogeneous, which implies that the presence of carbon-emitting ones are higher. The marginal emissions are in this case relative to the gas combined-cycle, which generates an important part of the national electricity, especially during peak-hours. These data are recapitulated in Table 6.1, which indicates the amount of electricity generated and emissions released for each technology over the year 2016. Firstly, it is striking that the main source of contamination in Spain is coal whereas it is gas for France. Secondly, it is interesting to notice that, despite a very low average rate of emissions 68 gCO_2/kWh, France has a higher marginal rate than Spain, which will play an important role in the global carbon balance at national level.

Thanks to these data, the total amount of CO_2 saved thanks to the installation of a PV system can now be calculated. As previously explained in the example, in France, solar energy newly installed substitutes the fuel-based electricity. In these conditions, a 2 kWp installation, which is the optimum according to the previous

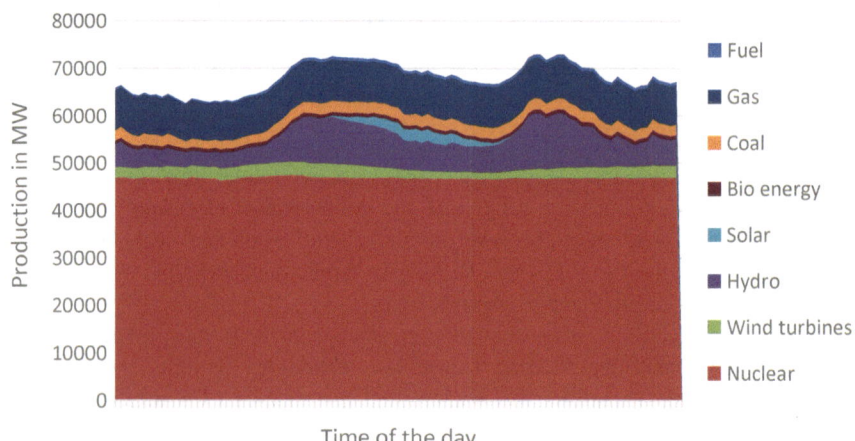

Fig. 6.1 Contribution of each technology for the French global production of electricity on December 14th 2016. *Source* RTE [1] and own elaboration

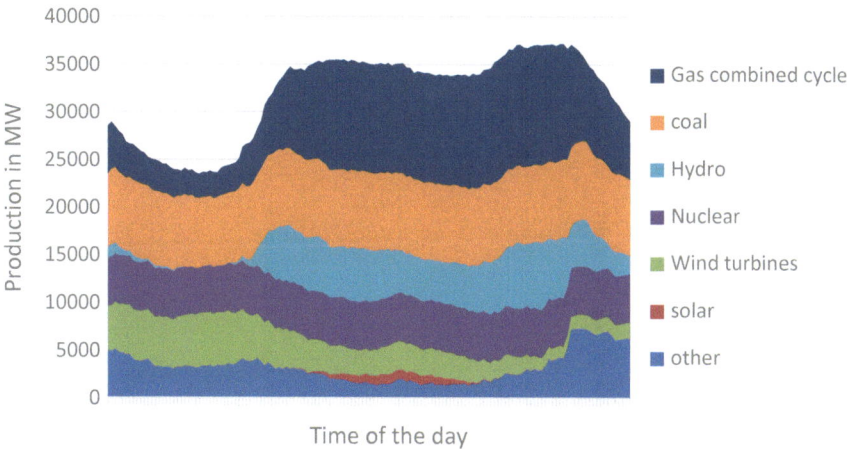

Fig. 6.2 Contribution of each technology for the Spanish global production of electricity on December 14th 2016. *Source* REE [2] and own elaboration

section and which annually produces 2050 kWh of electricity, can avoid the release of 1384 kg of CO_2 in the atmosphere each year. In Spain, this amount is a little lower (558 kg/year) since the solar energy compensates the generation from gas combined cycle and since the optimum capacity to maximize the profitability is only 1.5 kWp. In the last section of this chapter, these quantities will be generalized at a national level and compared to the reduction objectives established by Paris agreement.

6.2 Solar Modules Production Process

The manufacturing process is the only source of carbon emissions for a solar system because once installed it is self-dependent, therefore it does not produce emissions and neither requires energy supplying. Nevertheless, this process is quite complex and necessitates energy expenditures. This section aims at explaining the main steps of the process so that we can later better understand the energy and emissions involved. The following description is based on Stoppato's work (2008) [3] which deals with life cycle assessment of photovoltaic modules.

The main raw material to make PV modules is silica, a mineral which is very abundant on Earth but that requires a lot of transformation before being usable in this context. Stoppato's contribution (2008) [3] presents the 9 steps of the whole manufacturing process, where the four first stages are just about silica's treatment. The first extraction is made out of sand and is a process which is not very energy-intensive. Then the silica is transformed into silicon according to the chemical equation:

Table 6.1 Yearly contribution of each technology to the generation of electricity and the emissions that are linked with

France	Generation (MWh)	Emission (Mt)	Emission rate (gCO$_2$/kWh)	Spain	Generation (MWh)	Emission (Mt)	Emission rate (gCO$_2$/kWh)
Fuel	2,388,776	1.6	673	Fuel	0	0	
Gas	40,162,175	18.6	463	Gas	31,500,334	12	370
Coal	10,053,606	9.6	956	Coal	33,952,473	32	950
Nuclear	387,772,845	0	0	Nuclear	55,196,787	0	0
Hydro	67,494,269	0	0	Hydro	37,305,723	0	0
Solar	7,668,289	0	0	Solar	13,05,191	0	0
Wind turbines	24,630,201	0	0	Wind turbines	51,318,525	0	0
Other	8,296,612	8.1[a]	983	Other	33,450,977	0	0
Total	548,466,772	37.9	68	Total	255,930,011	44	163

Sources RTE [1], REE [2] (2016) and own elaboration

[a]These emissions are relative to the utilization of bio-energy

$$SiO_2 + 2C \rightarrow Si + 2CO.$$

At the end of this transformation the silicon is about 98% pure which is not enough for solar cells. This is why a second transformation, this time into solar silicon which purity is between $1\text{--}10^{-3}$ and $1\text{--}10^{-6}$, is required. It consists of silicon hydrogenation in a fluid bed reactor at 500 °C and 3.5 MPa with a copper-based catalyst and a series of fractionated distillations eliminating impurities. This is the most energy-intensive step, as the heating needs much electric energy. It is responsible for 47% of the whole process energy consumption according to Stoppato, when other papers estimate it at up to 60%. Then, the solar-silicon is cut and modeled in shape of wafers, and the chemical attack constitutes the first phase of the solar cell itself production. A $KOH\text{--}NH_3$ solution is used to remove the damages on the wafer surface. After that, other chemical treatments are realized until obtaining the solar cells. The last stage is therefore to assemble these cells into a panel. The assembling necessitates the introduction of other raw materials which are taken into account in the carbon footprint.

In his paper, Stoppato evaluated the emissions relative to each phase of the process and finally estimated that the manufacturing of a panel rejects in the atmosphere the equivalent of 80 kg of CO_2. A panel has a capacity of 250 W, it means that for the 2 kWp installation we studied in the anterior section 8 panels are required. Consequently, 640 kg of CO_2 are emitted for the production of the whole system. Besides, we calculated that according to our model this system will allow to save the emission of 3.5 tons of CO_2 over its life time. Therefore, its carbon footprint is really positive as it produces less than 20% of the emissions it avoids.

We can also evaluate the ecological impact of a PV system by determining the quantity of equivalent CO_2 emitted for each kWh produced. This quantity simply corresponds to the 640 kg of CO_2 rejected during the manufacturing divided by all the kWh the system will produce during its 25 years of running. In our model, a 2 kWp installation in France generates 2050 kWh/year, which is 51,250 kWh over 25 years. It results that the installation has an average emission rate of 12.5 gCO_2/kWh. The same reasoning for Southern Spain (a 1.5 kWp installation annually producing 1510 kWh) gives us a very similar carbon footprint: 12.8 gCO_2/kWh. These values are really low, in the next section they will be compared to with what the literature says and to another way of calculation.

6.3   Energy Payback Time and Carbon Footprint

The energy payback time (EPBT) is the time necessary for a system to produce the same amount of energy as the quantity needed for its fabrication. It is an indicator very often used to characterize renewable energies. For the photovoltaic sector, scientifics agree on an EPBT in a range of 1–3 years depending on the type of solar modules (mono or polycristallin) and on the geographic position of the installation.

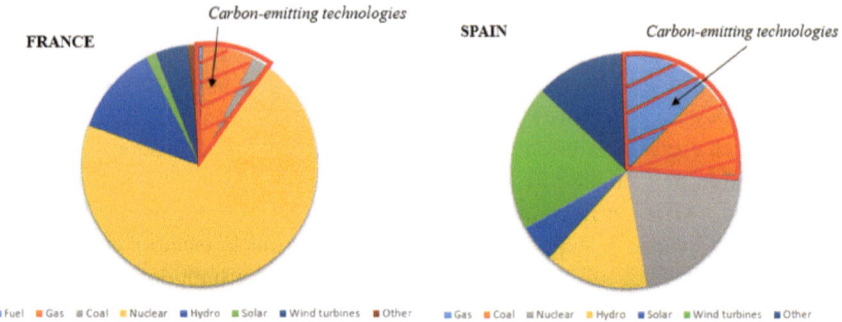

Fig. 6.3 Distribution between clear and carbon-emitting technologies for France and Spain. *Sources* RTE [1], REE [2] (2016) and own elaboration

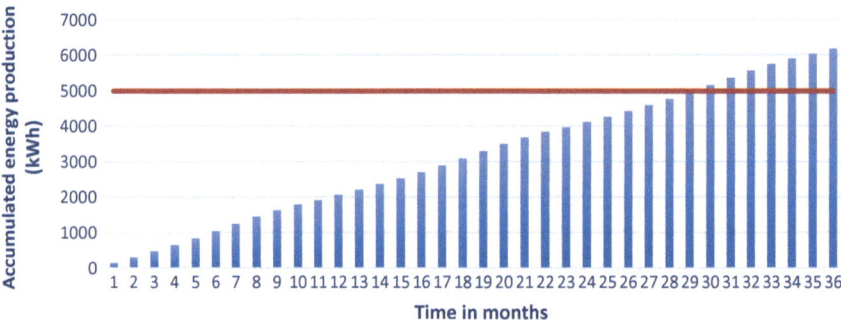

Fig. 6.4 EPBT for a 2 kWp system installed in Southern France. *Source* Own elaboration

In this section we are going to calculate the EPBT for our model. Unfortunately, it is quite complicated to obtain precise data on the emission related to the production process, this is why we will make several calculations based on different data source (Fig. 6.3).

Hespul is a French association working on PV energy and on the development of all renewable energies. According to one of their 2009 report, to install a 1 kWp solar system you need 2500 kWh of primary energy. Besides, with our model, a 2 kWp PV system in Marseille area produces around 2050 kWh of useful energy per year. Figure 6.4 illustrates the EPBT of such a system, we can see that the 5000 kWh are reached after 2.5 years which corresponds exactly to 10% of its lifespan.

If we consider that the value 2500 kWh to fabricate 1 kWp of solar modules is valid for every Mediterranean country we can determine the EPBT in function of the geographic area. Table 6.2 displays the results, without surprise the EPBT is lower in the most sunned areas, but in all the Mediterranean area it stays in a range between 2 and 3 years. It is worth reminding that our model is quite conservative because we consider only the useful energy produced and not all the electricity that

Table 6.2 EPBT in function of the geographic position

	Marseille	Madrid	Seville	Lisbon	Naples	Milan	Malta	Athens
EPBT (years)	2.5	2.75	2.5	2.75	3	3.25	2	3

Source Own elaboration

a PV system can produce during the year. If we were doing so, we would obtain EPBT even lower.

Another way of quantifying a solar system carbon footprint is to convert the 2500 kWh necessary to its fabrication in quantity of CO_2 rejected in the atmosphere and to compare it with what is saved over its lifespan.

To make the conversion between kWh and grammes of CO_2 we will assume that all the energy necessary to the fabrication is electricity. Then the carbon emissions relative to the production of electricity depend on the energy source chosen. As we saw in Sect. 5.1 the difference between electricity coming from nuclear or coming from coal is really significant. But we cannot know how much of the 2500 kWh comes from a source or another. Therefore we will consider the average quantity calculated at national level which is really different from one country to another. For France, we used data from RTE [4] and we calculated the average value picking out one day for each month of 2016. According to this calculation, we find that producing one kWh of electricity in France emits 68 g of CO_2 in the atmosphere. By comparison, in A. Stoppato's paper they evaluate the value for France at 82 g of CO_2. Using our value, manufacturing a 2 kWp solar system in France produces 340 kg of CO_2 which is equivalent to only 10% of the emissions avoided thanks to this system. This is a satisfactory result that is coherent with the EPBT estimated at 10% of the system's lifespan. For Spain we applied the same method of calculation, taking data from REE (Red eléctrica de España) and we obtained, without surprise, a higher result: 167 gCO_2/kWh. Thus, the manufacturing of the modules releases 626.25 kg of CO_2. The two last columns of Table 6.4 present the energy balance of the PV system relative to these cases: modules manufactured and installed in France or in Spain. They are the best scenarios, the carbon emissions avoided are respectively 100 and 22 times higher than the quantity emitted.

However, these results may be disconnected with reality since the main part of the solar modules installed in Europe are produced in China where the carbon emissions due to electricity generation are way higher (though official data do not exist, it is estimated that approximately 80% of European PV systems are imported from China). Cui-Mei and Quan-Sheng [5] carried out an extensive study on China's carbon emissions. China generates 69% of its electricity from coal, this is why its rate emission is so high. Furthermore, emissions are very different in function of the Province, they are included in a range between 300 and 1050 gCO_2/kWh. The average value at national level is 750 gCO_2/kWh. As a result, the fabrication of a 2 kWp PV installation in China generates 3.74 tons of CO_2. The two first columns of Table 6.3 figure this scenario in which the positive impact in the environment is slightly lower but stays completely satisfying.

Table 6.3 Energy balance in function of different fabrication and installation locations

Place of fabrication	China		Europe[a]		Spain	France
Place of installation	Spain	France	Spain	France	Spain	France
Emissions avoided (kg of CO_2)	−13,900	−34,600	−13,900	−34,600	−13,900	−34,600
Emissions generated by the manufacturing process (kg of CO_2)	2805	3740	1987.5	2650	626.25	340
Balance	−11,095	−30,860	−11,912.5	−31,950	−13,173.75	−34,260

Source Own elaboration

[a]Average emissions rate based on data from Stopatto (2006) [3] for the following countries Belgium, Czech Republic, Denmark, Finland, Germany, Greece, Hungary, Ireland, Italy, Luxembourg, the Netherlands, Portugal, Spain, United Kingdom

The last scenario considered in Table 6.3 is the fabrication in Europe (using an average value for the carbon emissions, 530 gCO_2/kWh[3]) and the installation still in France or in Spain. This case's results stand between the two previous one with emissions avoided again way more important than emissions generated.

These different scenarios raise the question of the relevance of realizing energy balances at a national level. Indeed, for Europeans countries installing PV systems on their territory, when it comes to make an energy balance, they take into account only the carbon that was not released thanks to this installation but not the emissions relative to the fabrication process since it was done abroad. By doing this they may obtain energy balance falsely positive, which can be a problem if it is done on a large scale. Nowadays, society is so globalized and international flows so developed than doing national carbon balances is most of the time not pertinent. Life cycle assessments have to be done at a global level to be more significant.

Considering these quantities of CO_2 produced by the fabrication process we can evaluate the carbon footprint of the system in each scenario. Results are displayed in Table 6.4, where also appear the carbon footprint found in the anterior section according to Stopatto's statement. Regardless of the method employed, we obtain that the carbon footprint of a PV system ranges between 8 and 80 gCO_2/kWh. These values are coherent (even if slightly superior) with those presented in the literature that go from 12 to 68 gCO_2/kWh (see Table 6.5). Nevertheless, obtaining such a wide range of values for a unique indicator can be a little unsatisfying. Indeed, promoting a green energy saying it releases only 10 g of carbon for each kWh generated is quite different from saying it produces 8 times more emissions. This is why additional explanations on these results are necessary.

The first thing to note is that the fabrication place influences much more the ecological impact that the PV system will have than its final location. Indeed, even

---

[3]Average emissions rate based on data from Stopatto (2006) [3] for the following countries Belgium, Czech Republic, Denmark, Finland, Germany, Greece, Hungary, Ireland, Italy, Luxembourg, the Netherlands, Portugal, Spain, United Kingdom.

Table 6.4 Carbon footprint in gCO_2/kWh, two ways of calculations and different scenarios considered

Place of fabrication	China		Europe		Spain	France
Place of installation	Spain	France	Spain	France	Spain	France
Hespul	83.1	83.3	58.9	58.9	18.6	7.6
Stopatto A.					14.1	13.9

Source Own elaboration

Table 6.5 Carbon footprint of PV energy according to several authors

	Carbon footprint (gCO_2/kWh)	Comments
Stylos and Koroneos (2014) [4]	12–55	Considerations about efficiency improvement and changes in the raw materials
Mulvaney [6]	32	System installed in Europe
	68	System installed in China
Parliamentary office of science and technology (2006) [7]	35	System installed in Southern Europe

Source own elaboration

if only France and Spain are here considered as places of installation, they are two countries with different solar irradiation and, above all, with distinct electricity generation systems, but the results for both cases are almost the same (see two first columns of Table 6.3).

The lowest value (7.6 gCO_2/kWh) is obtained considering that all the energy required for the manufacturing process of the solar modules is electricity and that this process takes place in France. Unfortunately, this theoretical result may not be very representative of the reality for several reasons. On one hand, very few PV systems running in Europe have been manufactured there. Though official data do not exist, it is estimated that approximately 80% of them are imported from China where the contamination and the carbon emissions are much higher. On the other hand, the hypothesis on the energy necessary being only electricity is quite optimistic, it could happen in an optimized system of production but it is not the case in the majority of them. Consequently, a carbon footprint of 7.6 gCO_2/kWh for solar energy is theoretically possible but the PV systems that are commonly used cannot reach such a low rate of emissions.

Then, if our results are slightly superior to what the literature generally says, it is because our model is conservative (due to the absence of energy storage, to the fact that no electricity is sold back to the grid and that only the useful energy is considered). But it means that these values (80 and 60 gCO_2/kWh respectively) are representative of a certain reality and could for instance be used in a pessimistic scenario for the evaluation of emissions in a larger context.

Table 6.6 Carbon footprint (in gCO_2/kWh) comparison for renewable energies

	Hydro-power energy	Wind turbines	Photovoltaic energy
Evans, Strezov, Evans [8]	41	25	90
Jacobson (2009) [9]	17–22	2.8–7.4	19–59
Mulvaney [6]	12	5	35
Pehnt [10]	10–13	9–11	104

Source Own elaboration

In any case, the emissions relative to solar energy are much lower than the one corresponding to fossil energies (except nuclear energy which is a special case). However, it is interesting to see how photovoltaic ranges among other renewable energies. The comparison is here limited to the wind-power and the hydraulic energies, which are often considered as the direct rivals of PV. The comparison with the nuclear would not be relevant since too many other factors would have to be integrated (storage of the nuclear waste, risks of explosions, ethical issues...). Table 6.6 displays the results of four authors who have carried out a comparative work on the carbon footprint of renewable energies. Two relevant points can be noticed. The first one is that in every case, PV energy has a carbon footprint way higher than the two other sources. The second one is that here again the proposed values vary from one to five in function of the paper. This illustrates the complexity of dealing with life cycle assessment, the large number of factors involved and the numerous ways of calculation existing both leading to a wide range of results. Nevertheless, determining the carbon footprint of an energy source remains a great way of measuring its impact on the environment and is an excellent tool to evaluate the emissions at a national or international level. Therefore, it can be very helpful for planning and reaching carbon-reduction objectives. This is why in the next section we will estimate to what extent the development of PV energy can help Spain and France to reach their objectives of reduction imposed by Paris agreement.

6.4   Contribution to Paris Agreement

The 21st conference of the Parties of the UNFCCC[4] in Paris in November 2015 was very eagerly awaited by all people involved in the protection of the environment or in any green movement. A lot of expectations were placed in the outcomes of this meeting between the highest world leaders. After weeks of debate and negotiations, the Paris agreement was finally adopted by consensus on December 12th 2015 and opened for signature on April 22th 2016. By December 2016, 194 UNFCCC members have signed the treaty, which is now applicable. The objective of this agreement is to limit the global warming at only 2 °C over pre-industrial levels.

---

[4]United Nations Framework Convention on Climate Change.

This is an ambitious goal as, according to specialists, it would require a diminution of 17 Gteq CO_2 of the global emissions to be reached. To meet this objective, a common effort must be realized, but each party of the treaty has also made an individual commitment of carbon emissions reduction called Intended Nationally Determined Contribution (INDC). The European Union has submitted a shared INDC which targets at least a 40% domestic reduction in greenhouse gas emissions by 2030 compared to 1990 and up to a 75% reduction by 2050.

In this paragraph, cases of France and Spain will be studied as they have been the common thread of this chapter. According to the Kyoto Protocol reference manual published by the UNFCCC, the initial 1990 amount of emissions on which are based the reduction objectives are 2820 and 1666 Gteq CO_2 for France and Spain respectively.

At the end of December 2016, the French Minister of Ecology, Sustainable Development and Energy officially published their low carbon strategy [11] in which they explain how they will reach the targets imposed by the Paris agreement. The emissions must be annually reduced by 9–10 Mteq CO_2 over the next 35 years. In this document, the contribution of each sector to the global emissions is detailed (see Fig. 6.5). The energy sector, in which we are interested here, is responsible for 12% of all the GHG emissions. Therefore, the objective is to modify the way of producing energy so as to emit between 1.08 and 1.2 Mteq CO_2 less each year. The point here is to know how much of this could be reached just by installing PV systems such as the one described in this paper.

As it was previously explained, one PV system installed in France helps to reduce the emissions relative to the fuel-based generation of electricity. But, if

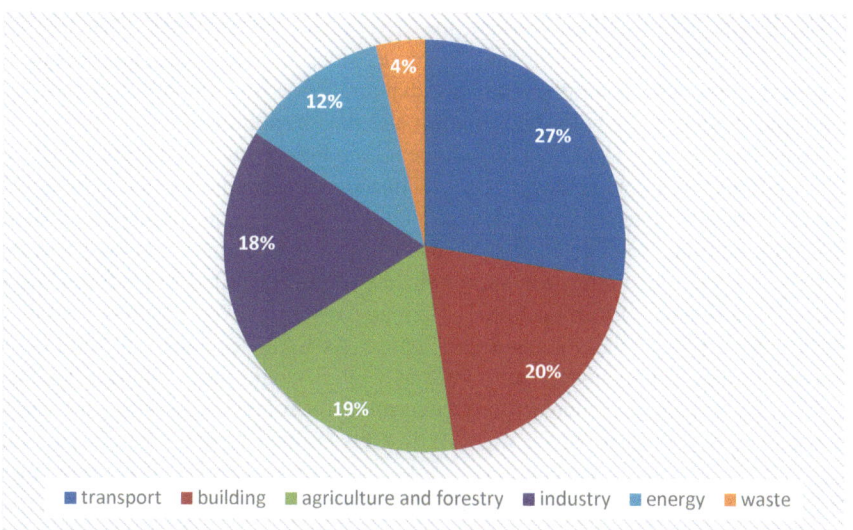

Fig. 6.5 Contribution of each sector to the global carbon emissions for France *Source* French low carbon strategy [11] and own elaboration

many PV systems are mounted we could hope to cancel all the emissions corresponding to the utilization of fuel and also the emissions due to the following technologies entering in the electricity generation. In France, the next two technologies to enter are the gas and the coal, and the rest of the electrical demand is fulfilled with zero emission sources (nuclear and renewable energies). Consequently, the maximum amount of electricity which is produced by carbon emitting sources is approximately 9000 MW (value reach during peak hours in winter), of which 300 MW are corresponding to fuel, 1800 MW to coal and 6900 MW to gas. To fulfill this demand only with 2 kWp domestic solar system, it would be necessary to equip 4.5 million houses. Given that there are 27 millions of dwellings in the country and that 40% of them correspond to houses and not apartments (data from the INSEE[5]) it is something feasible. Besides, installing PV systems in more than 4.5 million homes would not be interesting in terms of carbon emission considerations. Indeed, this investment is enough to compensate all the carbon-emitting technology in the generation of electricity, and if more solar modules were mounted they would substitute nuclear or hydraulic generation, which already are zero emission energies (only carbon emissions are taking into account here).

In this case, 150,000 systems would be installed to compensate the fuel utilization, annually avoiding the release of 0.207 $MtCO_2$; 3.45 millions would stand for the compensation of gas, corresponding to a 3.28 $MtCO_2$ reduction; and 900 000 for the coal, reducing 1.76 $MtCO_2$ more the emissions. In total, 5.24 $MtCO_2$ would not be released in the atmosphere each year. The reduction objectives could therefore be reached only with these new installations during the five first years of the project. Besides, the total amount of the investment would be worth 12.2 billion euros.

Finally, in its low carbon strategy, the government contemplates increasing the taxes on fossil energies more and more in order to encourage investment in renewable ones. This is another evidence that a massive investment in solar energy would be profitable for both the economy and the environment.

Spain has the same reduction commitment than France but has not published yet its long-term reduction strategy, we will hence compare the reduction obtained thanks to the PV system to the global reduction objective. The country has a share of carbon-emitting sources of energy much higher than France, the coal and the combined cycle accounting for approximately 7000 and 8000 MW during peak hours. As a result, 10 million homes equipped with 1.5 kWp PV systems could replace these two contaminating technologies. As Spain contains 25 millions of households and of which 40% are houses as well, it would mean taking advantage of the maximum domestic capacity of the territory for solar energy. This investment would allow to compensate for the 6.45 $MtCO_2$ emitted by the coal activity and 3.07 $MtCO_2$ relative to the combined cycle (their respective rate of emissions are 950 and 370 gCO_2/kWh). In total, 9.52 $MtCO_2$ would be saved each year. This

---

[5]Institut National de la Statistique et des Etudes Economiques.

amount represents only 0.57% of the 1990 level of emissions, but yet it would be a significant progress for the energy sector. Moreover, this project would require an initial investment of 24.5 billion euros and, thanks to its profitability, it would generate an annual revenue that could be re-injected in the market.

References

1. Eco2mix, data base of RTE. http://www.rte-france.com/fr/eco2mix/donnees-en-energie
2. Online data base of REE. https://demanda.ree.es/movil/peninsula/demanda/tablas/2016-01-14/3
3. Stoppato A (2008) Life cycle assessment of photovoltaic electricity generation. Energy 33:224–232
4. Stylos N, Koroneos C (2014) Carbon footprint of polycrystalline photovoltaic systems. J Clean Prod 64:639–645
5. Cui-Meil MA, Quan-Sheng GE (2014) Method for calculating CO_2 emissions from the power sector at the provincial level in China. Adv Climate Change Res 5(2):92–99
6. Mulvaney D (2014) IEEE spectrum, 13th Nov 2014
7. Carbon footprint of electricity generation (2006) Postnote . https://www.parliament.uk/documents/post/postpn268.pdf
8. Evans A, Strezov V, Evans T (2009) Assessment of sustainability indicators for renewable energy technologies. Renew Sustain Energy Rev 13(5):1082–1088
9. Jacobson MZ (2009) Review of solutions to global warming, air pollution, and energy security. Energy Environ Sci 2:148–173
10. Pehnt M, (2006) Dynamic life cycle assessment (LCA) of renewable energy technologies. Renew Energy 31(1):55–71
11. French low carbon strategy. http://unfccc.int/files/mfc2013/application/pdf/fr_snbc_strategy.pdf

Chapter 7
Conclusions

In this project, an extensive and multidisciplinary study on the residential use of solar energy was carried out. After the presentation of the chosen model, the focus was initially put on the economic issue. The model consists in a basic system that does not include energy stockage or resale to the grid. The first objective was to determine the optimum capacity particulars have to install to maximize their benefits. A higher capacity generates more energy but as the totality of this production cannot be consumed (because of the very distinct profiles of the load curve and the solar production parabola) it might be more profitable to choose a lower capacity, which would be cheaper and which would limit the losses of the system. The optimization was done using the typical financial indicators (NPV, IRR and Payback) and was conducted in different countries of the Mediterranean area.

Table 7.1 sums up the main results that are quite homogeneous for all the countries. The ratio NPV/investment is between 0.4 and 0.6 and the IRR ranges from 3 to 4%. There are two exceptions: Greece, whose relatively low amount of useful energy does not compensate the cheap electricity and where the profitability is therefore lower; and Malta where the domestic tariff of electricity is really expensive and consequently installing PV systems is economically very interesting.

Then, the second objective was to determine, according to this model, when grid parity for solar energy would be reach. For this the LCOE was calculated for each country and compared to their respective marginal cost of electricity. In Malta the LCOE is already lower than the domestic price of electricity, it means that for the clients submitted to this tariff (mainly second homes) it is cheaper to produce their own energy with solar modules than to buy it from the grid. For all the other areas the LCOE is still higher than the retail electricity price. According to the exponential projection model of Biondi and Moretto (2014), based on the learning and growth rates of solar energy, we can determine when the LCOE will meet the end user price, that is to say when grid parity will be reach. The calculations reveal that grid parity will take between two and five years to arrive, except for Greece where it

© The Author(s), under exclusive licence to Springer Nature Switzerland AG 2019

Á. Arcos-Vargas and L. Riviere, *Grid Parity and Carbon Footprint*, SpringerBriefs in Energy, https://doi.org/10.1007/978-3-030-06064-0_7

Table 7.1 Summary of the different results

	Seville	Marseille	Lisbon	Malta (residential)	Malta (domestic)	Athens	Naples
Investment (€)	2450	2700	2450	2450	2450	2200	2450
Optimum capacity (kWp)	1.5	2	1.5	1.5	1.5	1	1.5
Annual total production (kWh)	2540	3000	2430	2520	2520	2300	2270
Useful energy (kWh)	1510	2050	1380	1800	1800	1300	1245
NPV/investment	0.60	0.49	0.40	0.50	1.11	0.26	0.53
IRR (%)	4.57	3.82	3.2	3.92	7.78	2.15	4.1
Payback (years)	14	15	15	14	10	19	14
Marginal cost of electricity (€/kWh)	0.162	0.122	0.161	0.13	0.167	0.147	0.19
LCOE (€/kWh)	0.176	0.136	0.187	0.143	0.143	0.199	0.207
Grid parity	2018	2020	2021	2019	Already there	2026	2019
EPBT (years)	2.5	2.5	2.75	2	2	3	3

Source Own elaboration

is expected to take around 10 years. Given that our model is quite conservative, these values are truly encouraging for the future of the solar market. Within only a few years, grid parity will have reach a large part of Europe and without doubt it will give a new boost to the sector.

After the economic issue, the ecological problematic was tackled. Indeed, the fabrication process of solar modules is not without consequence on the atmosphere since it is a complex process requiring energy and releasing carbon emissions. Thus, the aim was to evaluate the environmental impact of our PV system, using the EPBT and the carbon footprint as main tools.

The EPBT was calculated according to Hespul's statement which affirms that 2500 kWh of primary energy are necessary for the manufacturing of modules with a 1 kWp capacity. Knowing the energy produced each year by the system (taking into account only the useful energy) it is then simple to obtain the EPBT. The results show that it ranges between 2 and 3 years for all the countries considered. It represents approximately 10% of the equipment's lifetime, which means the energy debt is quickly reimbursed.

Two distinct methods were used to determine the carbon footprint. The first approach was based on Stopatto's report that says the fabrication of a 250 Wp module releases 80 kg of CO_2. The carbon footprint is then the total amount of CO_2

emitted divided by the total amount of kWh generated by the system over its whole lifetime. With this method, it is worth 14 gCO_2/kWh, which is a value quite low in comparison with what is found in the literature.

The second method consists in converting the 2500 kWh required for the manufacturing process into emissions of carbon (assuming all this energy is electricity). This technique allows to consider different scenarios regarding the places of fabrication and installation. Indeed, the main part of the PV system functioning in Europe are manufactured in China where the average emissions rate relative to the generation of electricity is much higher than in anyother place. Very different results are therefore obtained depending mainly on the place of fabrication, the lowest carbon footprint (8 gCO_2/kWh) corresponding to fabrication and installation in France. The highest one is for fabrication in China and is ten times higher (83 gCO_2/kWh). To give sense to these values a rapid comparison with other renewable energies was made. Actually, solar energy has a carbon footprint in average superior to hydro-power and wind turbines, which are often seen as its direct rivals. However, the emissions rate stays much lower than for fossil energies, for instance coal emits 950 gCO_2/kWh and fuel around 700 gCO_2/kWh. Consequently the development of the solar energy has a really positive impact on the environment. In the last section of this project we tried to quantify this impact by evaluing the quantity of carbon that could be saved if PV residential system were installed at a national level. For instance, in France, if 4.5 million homes were equipped with solar energy, it could compensate all the electricity that is currently generated from fuel, coal and gas. Therefore all the electricity of the country would be produced from zero-emission technologies (only carbon emissions are taken into account) and it would reduce the annual carbon emissions of 5.24 $MtCO_2$.

Finally, this study shows that the domestic use of solar energy is profitable in both economic and ecological terms, which promises a great future for the sector. Nevertheless, we have to mention the obstacles that are still limiting its development. First, the numerous and tedious administrative procedures necessary to legalize a residential solar equipment associated to its relatively low cost-effectiveness still can demotivate some people to take the plunge. Furthermore, the profitability highly depends on the retail price of electricity, which is a very volatile parameter. Consequently, an investment profitable today may become disadvantageous within only a few years and, thus, solar energy cannot be considered as a zero-risk investment. Then, if these barriers are overcame and a massive installation of PV systems is made, solutions would need to be found to adapt the grid to this new way of generating electricity. Indeed, all the renewable energies are inconstant source of production and it is then much more complex to meet the right adequation between production and demand. The investigation on this issue may be one of the most challenging of the energy sector for the next decades and the solutions will surely go through improving the efficiency and the monitoring of the energy storage.

Reference

1. Biondi T, Moretto M (2014) Solar grid parity dynamics in Italy: a real option approach. Energy 80:293–302. https://doi.org/10.1016/j.energy.2014.11.072